SIXTH EDITION

D0218100

WHEN ARE WE EVER GOING TO USE THIS STUFF?

COLLEGE MATHEMATICS FOR THE LIBERAL ARTS MAJOR

By Jim Matovina and Ronald Yates
College of Southern Nevada

cognella
academic publishing

Bassim Hamadeh, CEO and Publisher
Michael Simpson, Vice President of Acquisitions
Jamie Giganti, Senior Managing Editor
Jess Busch, Senior Graphic Designer
Kristina Stolte, Senior Field Acquisitions Editor
Alexa Lucido, Licensing Coordinator
Allie Kiekhofer, Interior Designer

First published in the United States of America in 2016 by Cognella, Inc.

Trademark Notice: Product or corporate names may be trademarks or registered trademarks, and are used only for identification and explanation without intent to infringe.

Cover image Copyright in the Public Domain.
    Copyright© Depositphotos/DLeonis.
    Copyright© Depositphotos/venimo.
    Copyright in the Public Domain.
    Copyright in the Public Domain.
    Copyright in the Public Domain.
    Copyright © Depositphotos/pupunkkop.
    Copyright in the Public Domain.
    Copyright © Depositphotos/logoff.

Copyright in the Public Domain.
Copyright © Dustin M. Ramsey (CC BY-SA 2.5) at https://commons.wikimedia.org/wiki/File:License_plate_shed_near_Parrsboro,_NS_-_08659.jpg.
Copyright © Depositphotos/Subbotina.
Copyright © Depositphotos/seregam.
Copyright in the Public Domain.
Copyright © Depositphotos/odze.

Printed in the United States of America

ISBN: 978-1-63487-250-8 (pbk) / 978-1-63487-251-5 (br)

www.cognella.com   800-200-3908

TABLE OF CONTENTS

CHAPTER 3

STATISTICS   117

CHAPTER 4

PROBABILITY 183

CHAPTER 5

GEOMETRY

247

CHAPTER 6

VOTING AND APPORTIONMENT   313

PREFACE

This book represents an accumulation of material over a couple decades. At the College of Southern Nevada, Professors Jim Matovina and Ronnie Yates have been teaching liberal arts math classes, both online and in the classroom, regularly since 1996. They have written and shared their materials with each other and their colleagues over the years, and those collaborations ultimately led to the creation of this book.

Acknowledgments

We thank the following individuals for their tireless and diligent efforts in reviewing and contributing to this book.

- Denny Burzynski
- Dennis Donohue
- Billy Duke
- Bill Frost
- Michael Greenwich, PhD
- Eric Hutchinson
- Joel Johnson, PhD

- Garry Knight
- Andrzej Lenard
- Alok Pandey
- Jonathan Pearsall
- Kala Sathappan
- Ingrid Stewart, PhD
- Patrick Villa

Special gratitude is extended to all the students and instructors who reported errors in the first several editions and suggested improvements. Most of those reports were anonymous, so specific names cannot be provided.

Biographical materials have been abridged from the corresponding entries found on Wikipedia (http://en.wikipedia.org), and the majority of the photographs and images throughout the text were acquired from public domain sites at Wikimedia Commons and Pixabay. Other images came from the database and tools in MS Word, or were actual photographs taken by the authors.

Once again, thanks.

Jim & Ronnie

To the Instructor

This work was originally created as a textbook for the College of Southern Nevada's (CSN's) MATH 120 course, which is commonly referred to as "Liberal Arts Math." It is fun to teach, and as a terminal course, it is typically taken by students who need just one math class to earn their degrees. Since it does not satisfy any prerequisites to other courses, we are given a fair amount of flexibility in what and how we teach, and that is reflected in the survey nature of the course.

In all the material, you should stress practical applications over symbolic manipulations, but that does not mean you should totally discount the algebra. Instead, help the students learn where, when, why, and how the mathematics will help them in their lives.

To the Student

A few years ago, a class was presented with the following problem:

> A farmer looks across his field and sees pigs and chickens. If he counts 42 heads and 106 feet, how many pigs are in the field?

One student, instead of answering the question, sent an e-mail to the instructor stating, "I feel this question is unfair and misleading because pigs do not have feet. They have hooves."

In this class, and in all math classes for that matter, we are not trying to trick you. Although you will, undoubtedly, be presented with some challenging questions, be assured they are there to make you apply and extend your thought processes. Don't waste time and effort looking for technicalities that might invalidate a question. Instead, take the questions at face value, and spend your time trying to solve them.

We completely understand students taking this course are often Liberal Arts majors and/or need only this one math class in order to graduate. Not only that, but, in many cases, students often wonder why they need to take this class in the first place. All of us—yes, including your instructor—have sat in a math class at one time or another and thought, "When are we ever going to use this stuff?" Well, this is the class where we tell you.

—Jim & Ronnie

1 CONSUMER MATH

Money. It controls nearly everything in the world. Those who have it are often seen as more powerful and influential than those who do not. So, having a basic understanding of how money grows and is (or should be) spent is essential.

Understanding percentages is key to understanding how money works. Budgets are based on different percents of available money being allocated in specific ways. Sale discounts and taxes are based on the percent of cost of the items at hand. Credit cards and loans operate on interest rates, which—you guessed it—are given in percents.

The single largest purchase most people will make in a lifetime is a home. A basic understanding of mortgages and finance charges will make you a much wiser homebuyer and could easily save you tens of thousands of dollars over a 20- to 30-year period.

We could go on, but you get the picture.

1.1 On the Shoulders of Giants (Biographies & Historical References)

Historical References in this Book

Throughout this book, you will find many historical references. These fascinating glimpses into the past are there to provide a bit of a human element. Essentially, if you can appreciate some of the people and aspects that led to the development of the mathematics at hand, you can better understand the reasoning behind its existence.

For Consumer Math …

As a signer of both the Declaration of Independence and the Constitution, Benjamin Franklin is considered one of the Founding Fathers of the United States of America. His pervasive influence in the early history of the U.S. has led to his being jocularly called "the only President of the United States who was never President of the United States.[1]" Franklin's likeness is ubiquitous, and, primarily due to his keen understanding of the power of compound interest and constant advocacy for paper currency, has adorned all the variations of American $100 bills since 1928.

The Dow Jones Industrial Average (DJIA) is one of several stock market indices; it was created by nineteenth-century *Wall Street Journal* editor Charles Dow. It is an index that shows how certain stocks have traded. Dow compiled the index to gauge the performance of the industrial sector of the American stock market and, thus, the nation's economy.

On December, 10, 2008, Bernard "Bernie" Madoff, a former chairman of the NASDAQ Stock Market, allegedly told his sons the asset management arm of his firm was a massive **Ponzi scheme**—as he put it, "one big lie." The following day, he was arrested and charged with a single count of securities fraud, but one that accused him of milking his investors out of $50 billion. The 71-year-old Madoff was eventually sentenced to 150 years in prison.

---

1   Firesign Theater quote

Benjamin Franklin

Throughout his career, **Benjamin Franklin** was an advocate for paper money. He published *A Modest Enquiry into the Nature and Necessity of a Paper Currency* in 1729, and even printed money using his own press. In 1736, he printed a new currency for New Jersey based on innovative anti-counterfeiting techniques, which he had devised. Franklin was also influential in the more restrained and thus successful monetary experiments in the Middle Colonies, which stopped deflation without causing excessive inflation.

In 1785 a French mathematician wrote a parody of Franklin's *Poor Richard's Almanack* called *Fortunate Richard*. Mocking the un-bearable spirit of American optimism represented by Franklin, the Frenchman wrote that Fortunate Richard left a small sum of money in his will to be used only after it had collected interest for 500 years. Franklin, who was 79 years old at the time, wrote to the Frenchman, thanking him for a great idea and telling him that he had decided to leave a bequest of 1,000 pounds (about $4,400 at the time) each to his native Boston and his adopted Philadelphia. By 1940, more than $2,000,000 had accumulated in Franklin's Philadelphia trust, which would eventually loan the money to local residents. In fact, from 1940 to 1990, the money was used mostly for mortgage loans. When the trust came due, Philadelphia decided to spend it on scholarships for local high school students. Franklin's Boston trust fund accumulated almost $5,000,000 during that same time, and was used to estab-lish a trade school that became the Franklin Institute of Boston.[2]

The Dow Jones Industrial Average (DJIA)

Charles Dow compiled his DJIA index to gauge the performance of the industrial sector of the American stock market. It is the second-oldest U.S. market index, after the Dow Jones Transportation Average, which Dow also created. When it was first published on May 26, 1896, the DJIA index stood at 40.94. It was computed as a direct average by adding up stock prices of its components and dividing by the number of stocks in the index. The DJIA averaged a gain of 5.3% compounded annually for the 20th century; a record Warren Buffett called "a wonderful century" when he calculated that, to achieve that return again, the index would need to reach nearly 2,000,000 by 2100. Many of the biggest percentage price moves in the DJIA occurred early in its history, as the nascent industrial economy matured. The index hit its all-time low of 28.48 during the summer of 1896.

---

2   Copyright © Wikimedia Foundation, Inc. (CC BY-SA 3.0) at http://en.wikipedia.org/wiki/Benjamin_Franklin.

The original DJIA was just that—the average of the prices of the stocks in the index. The current average is computed from the stock prices of 30 of the largest and most widely held public companies in the United States, but the divisor in the average is no longer the number of components in the index. Instead, the divisor, called the DJIA divisor, gets adjusted in case of splits, spinoffs, or similar structural changes to ensure that such events do not in themselves alter the numerical value of the DJIA. The initial divisor was the number of component companies, so that the DJIA was, at first, a simple arithmetic average. The present divisor, after many adjustments, is much less than 1, which means the DJIA, itself, is actually larger than the sum of the prices of the components. At the end of 2008, the value of the DJIA Divisor was 0.1255527090, and the updated value is regularly published in the *Wall Street Journal*.[3]

Charles Ponzi & His Scheme

A **Ponzi scheme** is a fraudulent investment operation that pays returns to investors from their own money or money paid by subsequent investors, rather than from profit. Without the significant information about the investment, only a few investors are initially tempted, and usually for small sums. After a short period (typically around 30 days), the investor receives the original capital plus a large return (usually 20% or more). At this point, the investor will have more incentive to put in additional money and, as word begins to spread, other investors grab the "opportunity" to participate, with the promise of extraordinary returns. The catch is that, at some point, the promoters will likely vanish, taking all the remaining investment money with them.[4]

The scheme is named after **Charles Ponzi**, who became notorious for using the technique after emigrating from Italy to the United States in 1903. Ponzi did not invent the scheme, but his operation took in so much money that it was the first to become known throughout the United States. Ponzi went from anonymity to being a well-known Boston millionaire in just six months after using such a scheme in 1920. He canvassed friends and associates to back his scheme, offering a 50% return on investment in 45 days. About 40,000 people invested about $15 million all together; in the end, only a third of that money was returned to them. People were mortgaging their homes and investing their life savings, and most chose to reinvest,

---

3   Copyright © Wikimedia Foundation, Inc. (CC BY-SA 3.0) at http://en.wikipedia.org/wiki/Dow_Jones_Industrial_Average.

4   Copyright © Wikimedia Foundation, Inc. (CC BY-SA 3.0) at http://en.wikipedia.org/wiki/Ponzi_scheme.

rather than take their profits. Ponzi was bringing in cash at a fantastic rate, but the simplest financial analysis would have shown that the operation was running at a large loss. As long as money kept flowing in, existing investors could be paid with the new money. In fact, new money was the only source Ponzi had to pay off those investors, as he made no effort to generate legitimate profits.

Ponzi lived luxuriously: he bought a mansion in Lexington, Massachusetts with air conditioning and a heated swimming pool (remember, this was the 1920s), and brought his mother from Italy in a first-class stateroom on an ocean liner. As newspaper stories began to cause a panic run on his Securities Exchange Company, Ponzi paid out $2 million in three days to a wild crowd that had gathered outside his office. He would even canvass the crowd, passing out coffee and donuts, all while cheerfully telling them they had nothing to worry about. Many people actually changed their minds and left their money with him.

On November 1, 1920, Ponzi pleaded guilty to a single count of mail fraud before Judge Clarence Hale, who declared before sentencing, "Here was a man with all the duties of seeking large money. He concocted a scheme, which, on his counsel's admission, did defraud men and women. It will not do to have the world understand that such a scheme as that can be carried out ... without receiving substantial punishment." Ponzi was sentenced to five years in federal prison, and after three and a half years, he was released to face 22 Massachusetts state charges of larceny. He was eventually released in 1934 following other indictments, and was deported back to Italy, since he hadn't gained American citizenship. His charismatic confidence had faded, and when he left the prison gates, an angry crowd met him. He told reporters before he left, "I went looking for trouble, and I found it."[5]

---

5   Copyright © Wikimedia Foundation, Inc. (CC BY-SA 3.0) at http://en.wikipedia.org/wiki/Charles_Ponzi.

1.2 For Sale (Percents, Markup, & Mark Down)

When finding a percent of an amount or an amount that is a percent of a quantity, most people know they either need to multiply or divide by the decimal value of the percent. Unfortunately, most people can't quite remember when to *multiply* and when to *divide*. Fortunately, we can reduce all of those problems to the same equation and let our algebra skills tell us what to do.

A Brief Review of Rounding

Before we dive into the material for this course, let's take a few minutes to review how to correctly round numbers.

Rounding always involves a specified place value, and we *must* have a place value to round to. Without a specified place value, it is incorrect to assume one out of convenience.

> If there is no directive to round to a specified place value, it is incorrect to do so. After all, if no place value is mentioned, we cannot assume one out of convenience.

The formal procedure for the traditional rounding of a number is as follows.

1. First, determine the round-off digit, which is the digit in the specified place value column.

2. If the first digit to the right of the round-off digit is less than 5, do not change the round-off digit, but delete all the remaining digits to its right. If you are rounding to a whole number, such as tens or hundreds, all the digits between the round-off digit and the decimal point should become zeros, and no digits will appear after the decimal point.

3. If the first digit to the right of the round-off digit is 5 or more, increase the round-off digit by 1, and delete all the remaining digits to its right. Again, if you are rounding to a non-decimal number, such as tens or hundreds, all the digits between the round-off digit and the decimal point should become zeros, and no digits will appear after the decimal point.

4. For decimals, double-check to make sure the right-most digit of the decimal falls in the place value column to which you were directed to round and there are no other digits to its right.

Don't Get Trapped Memorizing Instructions

We cannot let ourselves get trapped in a process of trying to memorize a set of instructions. Yes, if followed correctly, instructions can be quite valuable, but when we blindly try to stick to a formal, rigid process, we often lose sight of logic and reason. For many students, rounding numbers is a classic example of this.

If, for a second, we set aside the traditional process outlined on the previous page and understand that "rounding a number to the tenth" can be thought of as "identifying the number ending in the tenths place that is closest to the given number," we can gain a better understanding of what it means to round.

For example, if we are tasked with rounding 14.68 to the nearest tenth, we could begin by identifying the consecutive numbers ending in the tenths place (remember, for decimals, the right-most digit of the answer must fall in the stated place value) that are immediately less than and immediately greater than 14.68. Those two numbers are 14.6 and 14.7. Then, we can simply ask ourselves, "Is 14.68 closer to 14.6 or 14.7?"

If we look at a number line…

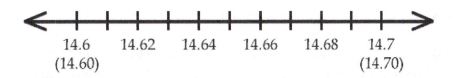

Treating 14.6 and 14.7 as 14.60 and 14.70, respectively, makes it easy to see 14.68 is closer to 14.7, than it is to 14.6. The important point to remember is the last digit of our rounded answer *must* fall in the place value to which we are rounding. In other words, the answer needs to be 14.7 and not 14.70.

EXAMPLE 1: Round 103.4736999 to the nearest tenth, hundredth, and then hundred.

For the tenth, begin by recognizing the four is in the tenths place. The seven immediately to its right indicates we are to change the four to a five, and remove the rest of the digits. Thus, rounded to the tenth, the value is 103.5.

For the hundredth, begin by recognizing the seven is in the hundredths place. The three immediately to its right indicates we are not to change the seven, and remove the rest of the digits. Thus, rounded to the hundredth, the value is 103.47.

For the hundred, begin by recognizing the one is in the hundreds place. The zero immediately to its right indicates we are not to change the one, and all the digits between the one and the decimal point are to become zeros. Thus, rounded to the hundred, the value is 100.

When to Round & When *Not* to Round

Throughout this book, and all of mathematics, for that matter, we will see many directives to "Round to the nearest hundredth (or other place value), when necessary." A very important part of that directive is the "when necessary" part. How are we supposed to know when rounding is necessary and when it isn't?

First, if we are given the absolute directive "round to the nearest hundredth"(or other place value) we MUST round to that specified place value. Keep in mind, when we are working with money, the underlying—but often unstated—directive is to round to the nearest cent.

Then we run into the "when necessary" scenario. This only applies to **non-terminating decimals**. If the decimal terminates, there usually is no need to round it. In fact, if we round it, our answer is less accurate than it could be. In other words, $1/32 = 0.03125$, exactly. If we round that decimal to the hundredth, to 0.03, the value is no longer equal to $1/32$. Sure, it's close, but it is NOT equal to $1/32$. 0.03 is actually $3/100$.

Non-terminating decimals are the ones to which we need to pay attention. If we dismiss the use of the "…" (called an "ellipsis"), it is impossible to write $1/3$ as a decimal. $0.33 = 33/100$, but $1/3 = 33/99$. $0.33333 = 33333/100000$, but $1/3 = 33333/99999$. 0.333333333333333 is closer, but as soon as we stop writing 3s, we no longer have exactly $1/3$. Thus, in order to save us from writing 3s indefinitely, it is necessary to round it to a specified place value.

Keep in mind; a calculator does not display an ellipsis. If you change $5/9$ to a decimal, the calculator may display the "final" digit as a 6: 0.555555556. Calculators typically round to the number of digits they can display on their screens. Some calculators may just truncate the decimal to the screen size, and others may actually hold an extra three to five digits of the decimal in memory without displaying them—this is the smarter version, since a greater amount of precision leads to a greater degree of accuracy.

If the decimal terminates within three or four decimal places, there may be no need to round. If it extends past four decimal places and we have a specified place value to which to round, then go ahead and do so. Remember, though, there must be a designated place value.

One of the few exceptions to this rule involves money. As mentioned earlier, if money is involved, we should always round to the nearest cent unless we are specifically told to round to a less precise value, such as the nearest dollar.

Identifying the Parts in a Percent Problem

In simple percent problems, there is a percent, a base, and an amount. The numeric value of the **percent**, p is easy to spot—it has a % immediately after it, and for calculation purposes, the percent must be changed to its decimal equivalent. The **base**, b is the initial quantity and is associated with the word "of." The **amount**, a is the part being compared with the initial quantity and is associated with the word "is."

EXAMPLE 2: Identify the amount, base, and percent in each of the following statements. If a specific value is unknown, use the variable p, b, and a for the percent, base, and amount, respectively. Do not solve the statements; just identify the parts.

    a. 65% of 820 is 533.

    b. What is 35% of 95?

    c. 50 is what percent of 250?

Answers

    a. Percent = 65%, Base = 820, Amount = 533

    b. Percent = 35%, Base = 95, Amount = a

    c. Percent = p, Base = 250, Amount = 50

Solving Percent Problems Using an Equation

Always begin by identifying the percent, base, and amount in the problem, and remember that, usually, one of them is unknown. If we are given a percent in the problem, change it to a decimal for the calculation. If we are finding a percent, the calculation will result in a decimal, , and we must change it to a percent for our answer.

Once we have identified the parts, we then need to recognize the basic form of the percent equation. An amount, a, is some percent, p, of a base, b. In other words, $a = pb$, where p is the decimal form of the percent.

When setting up the equation, if the a is the only unknown, find its value by simplifying the other side of the equation (by multiplying the values of p and b). If the unknown quantity is b, divide both sides of the equation by the value for p. Likewise, if the unknown quantity is p, divide both sides of the equation by the value for b.

EXAMPLE 3: What is 9% of 65?

The percent is 9%, so we will use $p = 0.09$. 65 is the base, so $b = 65$. Thus, the unknown quantity is a.

$a = 0.09(65) = 5.85$

EXAMPLE 4: 72 is what percent of 900?

900 is the base, and 72 is the amount. So, $b = 900$, and $a = 72$. The unknown quantity is the percent.

$72 = p(900)$

$72/900 = p$

So, $p = 0.08$, which is 8%.

General Applied Problems Involving Percents

To solve any application problem involving a percent, a good strategy is to rewrite the problem into the form "a is $p\%$ of b." Remember, we should always begin by clearly identifying the amount, base, and percent.

EXAMPLE 5: Of the 130 flights at Orange County Airport yesterday, only 105 of them were on time. What percent of the flights were on time? Round to the nearest tenth of a percent.

First, identify the base is 130 flights, and 105 is the amount we are comparing to the base. The percent is the unknown.

Then, we should recognize the simplified form of this question is "105 is what percent of 130?" This gives us the equation: $105 = p(130)$.

$105 = p(130)$

$105/130 = p = 0.8076923 \ldots = 80.76923 \ldots \%$

Rounding to the tenth of a percent, we state, "80.8% of the flights were on time."

In the above example, do notice the rounding was not done until the very end.

Sales Tax & Discounts

Sales taxes are computed as a percent of the cost of a taxable item. Then, when we find the dollar amount of the tax, we add it to the cost. **Discounts** are also computed as a percent of the original price of an item. For discounts, however, when we find the dollar amount of the discount, we deduct it from the original price. If both a discount and sales tax are involved in the same computation, we need to know which one gets applied first. In most cases, the sales tax gets computed on the discounted price. However, in some cases, we may get a discount, but will still be responsible for the taxes based on the original price—it all depends on the wording used in the situation. The latter of those two situations is usually referred to as a **rebate**. Often, rebates are given as dollar amounts, instead of percents.

Next, we need to make sure we answer the question being asked. If the question asks for the tax, we should *only* give the tax as our answer. Likewise, if the question asks for the discount, we should *only* give the amount of the discount as our answer. If the question asks for a sales price, then we need to be sure to subtract the discount from the posted price. If the question asks for the total cost of an object, we need to be sure to add the tax to the item's price.

Finally, when working with money, unless directed otherwise, we should always round to the nearest cent. Think about it. When we fill our gas tank and the total is $23.01, we have to pay that penny. Likewise, if your paycheck is for $543.13, you don't just get the $543.

> **Unless specifically directed to do otherwise, always round answers involving money to the nearest cent.**

EXAMPLE 6: Find the total cost of a $125 TV if the sales tax rate is 7.25%.

The base cost of the TV is $125, so $b = 125$. As a decimal, $p = 0.0725$.

$a = 0.0725(125) = 9.0625$. So, to the cent, the sales tax is $9.06.

That makes the total cost of the TV $125 + $9.06 = $134.06.

EXAMPLE 7: A pair of shoes is on sale for 20% off. If the original price of the shoes was $65, what is the sale price?

The base cost of the shoes is $65, so $b = 65$. $p = 0.20$.

$a = 0.2(65) = 13$. So, the discount is $13.

This makes the sale price of the shoes $65 − $13 = $52.

By the way, in the above example, 20% is a nice number to work with. We can quickly see 10% of $65 is $6.50. Twenty percent is twice as much, so by a quick inspection, 20% of $65 is 2 × $6.50, or $13. If we spot a quick inspection calculation like the one above, we definitely have a better understanding of the material at hand. If we cannot spot something like this, we can just resort to the computation shown in the example.

Percent Increase or Decrease

Another type of percent calculation we often see involves the change in percent. When studying information, it is often valuable to examine absolute changes in amounts over a period of time, as well as relative changes with respect to those amounts. The total amount something changes may be obvious, but examining the **change in percent** can often give a better overall picture of a situation or set of data.

Absolute Change is simply the difference between the beginning and ending amounts.

Percent Change, indicated with the notation Δ%, is found by the formula:

$$\Delta\% = \left(\frac{ending\ amount - beginning\ amount}{beginning\ amount} \right) \times 100$$

Keep in mind, if an amount is increasing, the Δ% will be positive. If the amount is decreasing, the Δ% will be negative.

EXAMPLE 8: Smith's Shoe Store has two salesmen, Al and Bob. Al sold 250 pairs of shoes in April and 322 pairs in May. Bob, who did not work as many hours as Al, sold 150 pairs of shoes in April and 198 pairs in May. Find the amount of increase in shoe sales for each salesman, and then find the percentage by which each salesman increased his sales, to the tenth of a percent.

For the change in the amounts, Al increased his sales by 72 pairs of shoes, and Bob increased his sales by 48 pairs.

For the percent of increase in the sales…

- For Al, Δ% = (322 − 250)/250 × 100 = 28.8%.

- For Bob, Δ% = (198 − 150)/150 × 100 = 32%.

In other words, even though he sold fewer pairs of shoes, Bob increased his rate of sales by a higher percent than Al.

EXAMPLE 9: Ronald Reagan's first term as President was 1981–85, and his second term was 1985–89. George H.W. Bush was President from 1989-93, and Bill Clinton was President from 1993-97, and again from 1997–2001.

The following table indicates the U.S. National Debt at the start of each of those presidential terms.

Year	Debt (in trillions)
1981	$0.9
1985	$1.6
1989	$2.6
1993	$4.0
1997	$5.3

During which presidential administration did the national debt increase the greatest amount? During which administration did the national debt increase at the greatest rate?

Answers:

George H.W. Bush increased the National Debt by $1.4 trillion.

Ronald Reagan's first term caused a 77.8% increase in the National Debt.

Section 1.2 Exercises

For Exercises #1 through #23, round bases and amounts to the nearest hundredth, and percents to the nearest tenth of a percent, as necessary.

1. What is 10% of 58?

2. What is 140% of 35?

3. 90 is what percent of 120?

4. 120 is what percent of 90?

5. 0.13 is what percent of 5?

6. 72 is 37% of what?

7. 45 is 112% of what?

8. 16 is 0.24% of what?

9. A student correctly answered 23 out of 30 questions on a quiz. What percent is this?

10. A basketball player made 13 out of 18 free throw attempts. What percent is this?

11. A recent survey showed 78% of people thought the President was doing a good job. If 5,000 people were surveyed, how many people approved of the President's job performance?

12. A bowl of cereal contains 43 grams of carbohydrates, which is 20% of the recommended daily value of a 2,000-calorie diet. What is the total number of grams of carbohydrates in that diet?

13. A bowl of cereal contains 190 milligrams of sodium, which is 8% of the recommended daily value in a 2,000-calorie diet. What is the total number of milligrams of sodium in that diet?

14. John's monthly net salary is $3,500, and his rent payment is $805. What percent of his net monthly salary goes toward rent?

15. Jane budgets 8% of her net salary for groceries. If she spent $300 on groceries last month, what is her monthly net salary?

16. Wade wants to buy a new computer priced at $1,200. If the sales tax rate in his area is 5.75%, what will be the amount of sales tax and total cost for this computer?

17. In the bookstore, a book is listed at $24.95. If the sales tax rate is 6.5%, what will be the amount of sales tax and the total cost of the book?

18. Mike has breakfast at a local restaurant and notices the subtotal for his food and coffee is $8.98. If the sales tax is 67¢, what is the sales tax rate?

19. Nancy wants to buy a sweater that is priced at $45. If the store is running a 40% off sale, what will be the amount of the discount and the sale price of the sweater?

20. Sue finds a CD she likes that normally costs $19.99. If the CD is labeled 75% off, and the sales tax rate is 5.5% (of the sale price), what will be the total cost of the CD?

21. To deal with slumping sales, Eric's boss cut his salary by 15%. If Eric was making $52,000 per year, what will be his new annual salary?

22. Last year, Ms. Rose had 24 students in her six-grade math class. This year she has 30 students. What is the percent change of her class size?

23. A new cell phone costs $40, and last year's model cost $45. What is the percent change in the cost of this phone?

24. Is it possible for one person to be 200% taller than another person? Why or why not?

25. Is it possible for one person to be 200% shorter than another person? Why or why not?

26. Because of losses by your employer, you agree to accept a temporary 10% pay cut, with the promise of getting a 10% raise in 6 months. Will the pay raise restore your original salary?

Answers to Section 1.2 Exercises

1. 5.8   2. 49   3. 75%   4. 133.3%

5. 2.6%   6. 194.59   7. 40.18   8. 6666.67

9. The student answered 76.7% of the questions correctly.

10. The player made 72.2% of his free throws.

11. 3,900 of the people surveyed approved of the President's job performance.

12. 215 carbohydrates are recommended in that diet.

13. 2,375 milligrams of sodium are recommended in that diet.

14. 23% of John's net salary goes toward his rent.

15. Jane's monthly net salary is $3,750.

16. The sales tax is $69, and the total cost of the computer is $1,269.

17. The sales tax is $1.62, and the total cost is $26.57.

18. The sales tax rate is 7.5%.

19. The discount is $18, and the sale price is $27.

20. The total cost of the CD is $5.28.

21. Eric's new annual salary will be $44,200.

22. Ms. Rose had a 25% increase in her class size.

23. The new phone costs 11.1% less than last year's model.

24. Sure. If a child is 2 feet tall, 200% of that is 4 feet. There are plenty of 2 feet children and 6 feet adults in the world.

25. Nope. Given the previous answer, it is tempting to say yes. However, 200% of 6 feet is 12 feet. Thus, to be 200% shorter than 6 feet, you would need to be −6 feet tall (6 − 12 = −6).

26. Nope. If your original salary was $10 per week, a 10% pay cut would make your salary $90 week. Then, a 10% raise would be 10% of the $90, not the $100. That means your new salary would be $99 per week.

1.3 The Most Powerful Force in the Universe (Simple & Compound Interest)

Simple vs. Compound Interest

Simple interest is just that: simple. It is calculated once during the loan period and is only computed upon the principal (money borrowed) of the loan or the investment. **Compound interest**, on the other hand, is calculated several times throughout the loan period. Furthermore, each time the interest is compounded, it is then added to the existing balance for the next compounding. That means we are actually paying (or earning) interest on interest!

Keeping it Simple

Interest is a percentage of a principal amount calculated for a specified period, usually a stated number of years. Thus, the formula for simple interest is **I = PRT**.

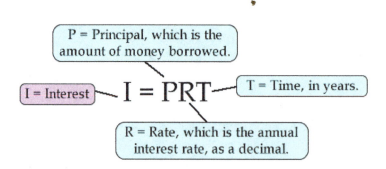

- **P** stands for the principal, which is the original amount invested.

- **R** is the annual interest rate. Be sure to convert the rate to its decimal form for use in the formula.

- The time, **T**, is also referred to as the term of the loan or investment and usually stated in terms of years.

It is possible to have the rate and time stated in terms of a time period other than years. If that is the case, make sure the two values are in agreement. That is, if we are working with a monthly rate, we need to state the time period in months, as well.

Also, remember, this formula gives us the amount of interest earned, so to find the total amount, A, of the loan or investment, we have to add that interest to the original principal, **A = P + I**.

EXAMPLE 1: Find the total amount returned for a 5-year investment of $2,000 with 8% simple interest.

First, find the interest. I = $2,000(0.08)(5) = $800. The total amount, A, returned to us is found by adding that interest to the original investment.

A = $2,000 + $800 = $2,800.

The amount returned at the end of the 5-year period will be $2,800.

Compounding the Situation

Legend has it, a Chinese emperor was so enamored with the inventor of the game of chess, he offered the guy one wish. The inventor asked for something along the lines of one grain of rice for the first square on the chessboard, two for the second square, four for the third, eight for the fourth, and so on—doubling for each of the 64 squares on the board. Thinking it was a modest request, the emperor agreed. Of course, upon finding out the inventor would receive $2^{64} - 1 = 18,446,744,073,709,551,615$ grains of rice (more than enough to cover the surface of the earth), he had the guy beheaded.

Perhaps that story is what inspired **Albert Einstein** to refer to compound interest as the "most powerful force in the universe."

The formula for the compounded amount is **A = P(1+r/n)^{nt}**.

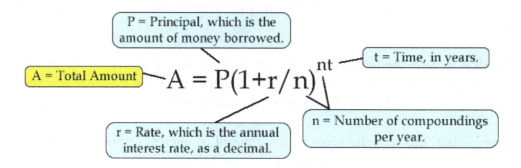

Don't fall into the trap of trying to memorize that formula. When we attempt to memorize something, we ultimately forget it or, even worse, recall it incorrectly. If we want to be able to recall a formula correctly, we need to understand the various components of it.

For the compounded amount formula, one of the most important things to realize is, unlike the formula for simple interest, the formula returns the total amount of the investment.

> The compounded amount formula returns the entire compounded amount, A. If we want to find only the interest, we need to subtract the original amount, $I = A - P$.

Just like in the simple interest formula, P stands for the principal, which is the original amount invested, and **t** stands for the time of the loan or investment stated in terms of years.

In the compounded amount formula, the interest rate, **r**, is also referred to as the **Annual Percentage Rate** (APR). Be sure to convert the APR to its decimal form for use in the formula.

Next, pay special attention to **n**, the number of compoundings per year. In most instances, this is described by using terms like monthly, annually, or quarterly. Make sure you know what each of those terms means.

Term:	Monthly	Quarterly	Semiannually	Annually
# of Times per Year:	12	4	2	1

Since **n** refers to the number of compoundings per year, **r/n** is the rate for each period. For example, if the APR is 15%, and the amount is compounded monthly, the monthly rate is $15\%/12 = 1.25\%$. Do keep in mind, though, we must still use the APR in the formula.

We add **1** to the monthly rate to account for the interest being added to the original amount. This is similar to a sales tax computation. If an item sells for $10, and the tax rate is 7%, the amount of the tax is $10(0.07) = $0.70, which would then be added to the original $10 to get a total cost of $10.70. Alternatively, we could compute it all in one step by multiplying $10(1+0.07) = $10(1.07) = $10.70. The **(1+r/n)** in the compound amount formula is exactly like the (1+0.07) in that tax calculation we did earlier.

The exponent in the formula, **nt**, is the total number of compoundings throughout the life of the investment. That is, if the amount is compounded monthly for 5 years, there will be a total of 12(5) = 60 compoundings.

EXAMPLE 2: Find the total amount due for $2,000 borrowed at 6% Annual Percentage Rate (APR) compounded quarterly for 3 years.

First, since quarterly implies four compoundings per year, **r/n** = 0.06/4 = 0.015. Then, add **1**, and raise that sum to the power that corresponds to the number of compoundings. In this case, 4 times a year for 3 years is 12. Finally, multiply by the amount borrowed, which is the principal.

$$A = \$2{,}000 \times (1 + 0.06/4)^{4 \cdot 3} = \$2{,}000 \times (1.015)^{12} = \$2{,}000 \times 1.195618 = \$2{,}391.24.$$

The total amount at the end of 3 years will be $2,391.24.

Using Your Calculator

In the previous example, we were faced with a multi-step calculation. Using a scientific calculator effectively can greatly expedite the process. For the exponentiation, $(1.015)^{12}$, we could literally multiply (1.015)(1.015)(1.015) … (1.015) twelve times, but it is clearly faster to use the exponentiation key on our calculator. Depending on the calculator, the exponentiation key may look like x^y or ∧. Thus, $(1.015)^{12}$ would be done on a scientific calculator as 1.015 x^y 12, and then, hit the = key. For more involved computations, we can use the parentheses keys. In all, we should never have to hit the = key more than once.

EXAMPLE 3: Use a calculator to compute $\$1,000(1 + 0.14/12)^{24}$.

Literally type: 1000 $\boxed{\times}$ $\boxed{(}$ 1 $\boxed{+}$ 0.14 $\boxed{\div}$ 12 $\boxed{)}$ $\boxed{x^y}$ 24 $\boxed{=}$

You should get 1,320.9871, which you will manually round to $\$1,320.99$.

EXAMPLE 4: On the day her son was born, Tricia invested $\$2,000$ for him at a guaranteed APR of 8%, compounded monthly. How much will that investment be worth when he turns 50? How much will it be worth if he leaves it alone until he turns 65?

On his 50th birthday, the investment is worth:
$A = \$2,000 \times (1 + 0.08/12)^{12 \cdot 50} = \$2,000 \times (1.00666667)^{600}$
$A = \$2,000 \times 53.878183179 = \$107,756.37$

On his 65th birthday, the investment is worth:
$A = \$2,000 \times (1 + 0.08/12)^{12 \cdot 65} = \$2,000 \times (1.00666667)^{780} = \$356,341.84$

EXAMPLE 5: If $\$1,500$ is invested at 5% APR compounded annually for 10 years, how much interest is earned?

The total amount, $A = \$1,500 \times (1 + 0.05/1)^{1 \cdot 10} = \$2,443.34$. Thus, the interest earned is found by subtracting the value of the original investment:
$\$2,443.34 - \$1,500 = \$943.34$.

Annual Percentage Yield

Often, when interest rates are quoted for a loan, we see two rates. One of them is the APR used in the compound interest calculation. The other is the **Annual Percentage Yield (APY)**, and is the simple interest rate that would give the same amount of interest as the APR over one year. Occasionally, this is called the **Effective Yield**, but we will stick with the APY.

Another way to describe the difference between the APR and the APY is by examining the calculations. Using the same principal, if we performed a compound interest calculation with the APR, we should see the same result if we performed a simple interest calculation with the APY, provided the calculations were done for a single year.

EXAMPLE 6: $1,000 is deposited into a Certificate of Deposit (CD) for one year, compounded monthly at an APR of 4%. Show that an APY of 4.074% would produce the same amount of interest over the same year.

Using the APR, $A = \$1,000(1+0.04/12)12 = \$1,040.74$. So, $I = \$40.74$

Using the APY, $I = \$1,000(0.04074)(1) = \$40.74$

Remember, when interest is compounded more than once a year, the true annual rate is higher than the quoted APR. The APY is the simple interest rate that takes into account all of the compoundings in a given year. Also, both the APR and APY are usually rounded to the nearest hundredth of a percent.

If we want to find the APY, we can use the formula, $APY = (1+APR/n)^n - 1$, where, as before, n is the number of compoundings per year. And, remember to use the decimal form of the APR.

EXAMPLE 7: Find the APY for a stated APR of 6.50% compounded quarterly. Round the answer to the nearest hundredth of a percent.

$APY = (1+0.065/4)^4 - 1 = (1.01625)^4 - 1 = 1.0666 - 1 = 0.0666 = 6.66\%$

Section 1.3 Exercises

1. Find the simple interest on $12,000 for 2 years at a rate of 6% per year.

2. Find the simple interest on $25,000 for 6 months at a rate of 8.5% per year.

3. How much time is needed for $1,800 to accumulate $360 in simple interest at a rate of 10%?

4. How much time is needed for $600 to accumulate $72 in simple interest at a rate of 4%?

5. If $4,300 yields $1,290 in simple interest over 6 years, what is the annual rate?

6. If $200 yields $45 in simple interest over 3 years, what is the annual rate?

7. How much principal is needed to accumulate $354.60 in simple interest at 9% for 4 years?

8. How much principal is needed to accumulate $65,625 in simple interest at 15% for 7 years?

9. If $500 yields $40 in simple interest over 2.5 years, what is the annual rate?

10. How much time is needed for $1,250 to accumulate $375 in simple interest at a rate of 5%?

11. Find the simple interest on $900 for 18 months at a rate of 9.5% per year.

12. If $420 yields $31 in simple interest over 30 months, what is the annual rate?

13. How much time is needed for $660 to accumulate $514.80 in simple interest at a rate of 12%?

14. Find the simple interest on $1,975 for 3 1/2 years at a rate of 7.2% per year.

15. Find the interest, I, and the total amount, A, for $825 invested for 10 years at 4% APR, compounded annually.

16. Find the interest, I, and the total amount, A, for $3,250 invested for 5 years at 2% APR, compounded annually.

17. Find the interest, I, and the total amount, A, for $75 invested for 6 years at 3% APR, compounded semiannually.

18. Find the interest, I, and the total amount, A, for $1,550 invested for 7 years at 5% APR, compounded semiannually.

19. Find the interest, I, and the total amount, A, for $625 invested for 12 years at 8% APR, compounded quarterly.

20. Find the interest, I, and the total amount, A, for $2,575 invested for 2 years at 4% APR, compounded quarterly.

21. Find the interest, I, and the total amount, A, for $1,995 invested for 6 years at 5% APR, compounded semiannually.

22. Find the interest, I, and the total amount, A, for $460 invested for 7 years at 6% APR, compounded quarterly.

23. To purchase a refrigerated showcase, Jamestown Florists borrowed $8,000 for 6 years. The simple interest was $4,046.40. What was the rate?

24. To purchase new equipment, Williams Brake Repair borrowed $4,500 at 9.5%. The company paid $1,282.50 in simple interest. What was the term of the loan?

25. John Doe earned $216 simple interest on a savings account at 8% over 2 years. How much was in the account?

26. Michelle is 25 and wants to retire at the age of 50. If she invests a $60,000 inheritance at 7% compounded semiannually, how much will she have when she turns 50?

27. In order to pay for college, Brooke's parents invest $20,000 in a bond that pays 8% interest compounded semiannually. How much money will there be in 18 years?

28. A couple sets aside $5,000 into a savings account that is compounded quarterly for 10 years at 9%. How much will the investment be worth at the end of 10 years?

29. Find the APY for a stated APR of 12% compounded monthly.

30. Find the APY for a stated APR of 8% compounded quarterly.

31. Find the APY for a stated APR of 9% compounded annually.

32. Find the APY for a stated APR of 4.5% compounded monthly.

33. When will the APR and the APY be equal?

Answers to Section 1.3 Exercises

1. $1,440

2. $1,062.50

3. 2 years

4. 3 years

5. 5%

6. 7.5%

7. $985

8. $62,500

9. 3.2%

10. 6 years

11. $128.25

12. 2.95%

13. 6.5 years (or 78 months)

14. $497.70

15. I = $396.20, A = $1,221.20

16. I = $338.26, A = $3,588.26

17. I = $14.67, A = $89.67

18. I = $640.11, A = $2,190.11

19. I = $991.92, A = $1,616.92

20. I = $213.36, A = $2,788.36

21. I = $688.05, A = $2,683.05

22. I = $237.92, A = $697.92

23. 8.43%

24. 3 years

25. $1,350

26. $335,095.61

27. $82,078.65

28. $12,175.94

29. 12.68%

30. 8.24%

31. 9%

32. 4.59%

33. The APY and the APR will be equal when the APR is compounded annually, which means the interest is simple interest.

1.4 Spending Money You Don't Have (Installment Buying & Credit Cards)

What is Installment Buying?

Installment buying is the process of purchasing something and paying for it at a later date. Do not confuse this with **layaway buying**, in which a customer makes monthly payments and then obtains the product after it is paid for in full. For the convenience of paying later, the consumer normally has to pay a finance charge. Also, the terms of any such agreement must, by law, be disclosed in writing at the beginning of the process. This law is the **Truth-in-Lending Act** (TILA). More information about the TILA can be found online in the Wikipedia entry at http://en.wikipedia.org/wiki/Truth_in_Lending_Act.

To simplify the calculations we will encounter in this section, we will not consider any applicable taxes that would normally be imposed on the indicated purchases. Conceptually, the calculations are the same with or without those taxes.

Simple Finance Charges

The **finance charge** is the amount of the purchase in excess of the selling price. Calculating the cost of financing is a very important, but straight-forward process.

EXAMPLE 1: Joel wants to purchase a TV for $1,075. Since he does not have the cash up front, he enters into an agreement to pay $35 a month for 36 months. Find the total cost and the finance charge for this purchase.

The total amount he will pay is $35 \times 36 = \$1,260$. Therefore, the finance charge would then be $1,260 - $1,075, or $185.

Monthly Payments

A critical component of installment buying is the calculation of the monthly payment. The tricky part of determining a monthly payment is realizing the last payment will usually be slightly more or slightly less than all the other monthly payments.

EXAMPLE 2: A recliner is for sale at a price of $450, but it may be purchased on an installment plan by paying $100 down, and then paying the balance plus 18% simple interest in 12 monthly payments. What would be the amount of each payment?

Using the simple interest formula, we will calculate the amount of interest that is to be paid: $I = PRT$, or $I = (\$350)(0.18)(1) = \$63$. Remember, the interest rate has to be in decimal form and the time is stated in years.

Adding that $63 to the balance of $350, the total amount of the payments will be $413. Since there will be 12 payments, each one should be $\$413/12 = \$34.416666\ldots$. However, since payments should be rounded to the cent, the first 11 payments would be 34.42, and the last payment would be the remaining amount of the purchase. $11 \times \$34.42 = \$378.62$, and $\$413 - \$378.62 = \$34.38$. So, the last payment on the purchase will be $34.38.

The total cost of the recliner would be the price of $450 plus the finance charge of $63, or $513 total.

EXAMPLE 3: Sarah buys a new mountain bike for $375 and pays for it over two years with 13.5% simple interest. What is her monthly payment?

To find the amount of interest she will pay, use the simple interest formula: $I = (\$375)(0.135)(2) = \$101.25$

This brings the total cost for the bike up to $476.25. Since she will be making 24 payments, $\$476.25/24 = \$19.84375$. That means the first 23 payments will be $19.84, and the last one will be $19.93.

Do note, in the two previous examples, the assumption is that no additional money is paid on top of the monthly payment. That is, if the payment happens to be $19.84, we are assuming the amount paid is exactly $19.84 and not a convenient $20.

Credit Card Finance Charges

The finance charge on a credit card is the simple interest on the average daily balance using a daily interest rate. In such a calculation, we also need to know how many days are in the billing cycle.

The **billing cycle** is the number of days between credit card statements, and usually corresponds to the number of days in a specific month, which will be 28 (29 in a leap year!), 30 or 31 days. We could pull out a calendar and count the days, but a faster way is to simply recognize the month that we are considering, and look to see if there are 30 or 31 days in it.

EXAMPLE 4: How many days are in the billing cycle that runs from June 15 through July 14?

This billing cycle crosses over the end of June. Since there are 30 days in June, there are 30 days in billing cycle.

The **Average Daily Balance** (ADB) is just that—the average of the daily balances for all the days in the billing cycle. We could find the ADB by finding the balance for every day in the billing cycle, adding them all together, and then dividing that sum by the number of days in the billing cycle. If, however, we are lucky enough to see the balance remain unchanged for stretches of several days at a time, the calculation can be a little quicker.

EXAMPLE 5: If a credit card has a balance of $20 for 21 days and then a balance of $40 for 9 days, what is the average daily balance for that 30-day period?

Rather than adding together $20 + $20 + $20 + $20 + … + $20 + $20 (21 times), we can simply multiply $20 × 21 to get $420. Likewise, for the 9-day stretch, the sum of those daily balances will be $40 × 9, or $360. Then, for the entire 30-day period, the sum of all 30 individual balances will be $780. Thus, the ADB for that 30-day period will be $780/30, which is $26.

The final piece of the credit card interest puzzle is the **daily interest rate**. The interest rate stated with a credit card is always an annual percentage rate (APR). Since there are 365 days in a year (366 in a leap year!), the daily interest rate is the APR/365.

Putting it all together … The monthly finance charge on a credit card is,

> **Monthly Finance Charge = (ADB) × (APR/365) × (# of Days in Billing Cycle)**

EXAMPLE 6: Let's say we have an average daily balance of $183.65 on a credit card with an APR of 16.9%, over a billing period of 31 days. What would the finance charge be?

Remember, we need to change the APR to decimal form …

$183.65 × (0.169/365) × (31) = $2.64 (to the nearest cent)

Further Credit Card Calculations

The balance on the Doe's credit card on May 12, their billing date, was $378.50. For the period ending June 11, they made the following transactions.

- May 13, Charge: Toys, $129.79

- May 15, Payment, $50.00

- May 18, Charge: Clothing, $135.85

- May 29, Charge: Gasoline, $37.63

a. Find the average daily balance for the billing period.

b. Find the finance charge that is due on June 12. Assume an APR of 15.6%.

c. Find the balance that is due on June 12.

SOLUTIONS

To answer these questions, or, for that matter, any question, our chances are greatly improved if we are organized and neat. First determine what it is we are trying to find. For question a) we are trying to find the average daily balance. Before we try to calculate it, can you define it?

QUESTION 1: What is an average daily balance?

Answer: The average daily balance is the average of all of the daily balances throughout the billing cycle. It is found by adding together the outstanding balance for each day and, then, dividing by the number of days in the cycle.

So, to do the calculation, we must begin by finding the balance for each day in the month. Fortunately, for a couple of long stretches, the balance remains unchanged. What about the billing date? Should it be included in the calculation?

QUESTION 2:   Should the billing date be included in the calculation?

Answer: The billing date signifies the beginning of the next billing cycle. In other words, for this problem, the issuance of the monthly statement (and, thus, the billing date) occurs on June 12. So, do not include June 12 in the calculations for this billing cycle—it will be the first day of the next billing cycle.

Here are the necessary calculations for this problem:

Dates	# of Days	Debits (Charges)	Credits (Payments)	Balance Due	# of Days × Balance
May 12	1			$378.50	$ 378.50
May 13–14	2	$129.79		$508.29	$ 1,016.58
May 15–17	3		$50.00	$458.29	$ 1,374.87
May 1–28	11	$135.85		$594.14	$ 6,535.54
May 2–June 11	14	$ 37.63		$631.77	$ 8,844.78
Totals:	31				$18,150.27

QUESTION 3:   How were the numbers in the "Balance Due" column calculated?

Answer: The Balance Due numbers are found by adjusting the outstanding balance by the indicated transaction—add purchases (Debits) and subtract payments (Credits). For example, going into May 18, the balance was $458.29. Then, add in the purchase of $135.85 to get the Balance Due of $594.14.

QUESTION 4:   How do we determine the number of days with the same balance?

Answer: The number of days can be determined two ways. We can make a list for the entire month, or we can simply subtract. For example, the balance changed on the 15th and again on the 18th. 18 – 15 = 3, so, for three days (the 15th, 16th, and 17th) the balance remained unchanged. Be careful when the billing cycle crosses into the next month. Since there is a May 31, the last stretch of days is 14 days. If this were April's bill, that last stretch would only contain 13 days.

QUESTION 5: At the end of the row beginning with May 15, how was the amount of $1,374.87 determined?

Answer: 3 days at a balance of $458.29. This gives us 3 × $458.29 = $1,374.87.

QUESTION 6: What are the numbers in the last row (31 and $18,150.27)?

Answer: The 31 is the number of days in the billing cycle. The $18,150.27 is the sum of the outstanding balances for every day in the billing cycle. Both numbers are found by adding together the numbers directly above them.

By the way, a nice little check in the middle of the calculations is to verify that the sum of the days in the "# of Days" column is, indeed, the correct number of days in the billing cycle. For this problem, as we noted earlier, since we cross over May 31, there are 31 days in this billing cycle. Since the sum of the days in the "# of Days" column is 31, we have accounted for every day in the cycle.

And, thus, to answer the three original questions ...

a. The average daily balance is $18,150.27/31 = $585.49

b. The finance charge is ($585.49) × (0.156/365) × (31) = $7.76

c. The balance due on June 12 is $631.77 + $7.76 = $639.53

ONE MORE QUESTION ...

QUESTION 7: Why was the finance charge added to the $631.77 and not $585.49?

Answer: The finance charge gets added to the final outstanding balance, not the average daily balance.

HOW ABOUT ANOTHER ... ?

The balance on the Smith's credit card on November 8, their billing date, was $812.96. For the period ending December 7, they made the following transactions.

- November 13, Charge: Toys, $231.10

- November 15, Payment, $500

- November 29, Charge: Clothing, $87.19

*A*ssuming they have an APR of 5.9%, find the balance that is due on December 8.

Here are the necessary calculations for this problem:

Dates	# of Days	Debits (Charges)	Credits (Payments)	Balance Due	# of Days × Balance
Nov. 8–12	5			$ 812.96	$ 4,064.80
Nov. 13–14	2	$231.10		$1,044.06	$ 2,088.12
Nov. 15–28	14		$500.00	$ 544.06	$ 7,616.84
Nov. 2–Dec. 7	9	$ 87.19		$ 631.25	$ 5,681.25
Totals	30				$19,451.01

- The average daily balance is $19,451.01/30 = $648.37.

- That makes the finance charge $648.37 × 0.059/365 × 30 = $3.14.

- Then, the balance forwarded is $631.25 + $3.14 = $634.39.

Credit Cards in General

Credit cards can be very useful. Unfortunately, they are also very dangerous. In the late 1980s credit card companies would target unsuspecting college students who were just a year or two away from their parents' watchful eyes. Many times, those students did not even have jobs or a regular income; the companies were issuing the credit cards to the students based solely on potential income. In turn, many college students would rack up thousands of dollars' worth of debt without even realizing the consequences. New credit card laws that went in to affect in February of 2010 specifically attempt to limit this practice with a few restrictions. Credit card companies are now banned from issuing cards to anyone under 21, unless they have adult co-signers on the accounts or can show proof they have enough income to repay the card debt. The companies must also stay at least 1,000 feet from college campuses if they are offering free pizza or other gifts to entice students to apply for credit cards.

Other highlights from the **Credit CARD Act of 2009** (CARD stands for **C**ard **A**ccountability **R**esponsibility and **D**isclosure) include limits on interest rate hikes and fees, allowing card holders at least 21 days to pay monthly bills, mandates on due dates and times, and clear information on the consequences of making only minimum payments each month. For that last one, the credit card companies must identify how long it would take to pay off the entire balance if users only made the minimum monthly payment.

If you are just starting out on your own, it is a good idea to avoid the overuse of credit cards. If you do not have the money, give a second thought to the purchase. Is it something you really need? Instead, try to think of a credit card as a cash substitute. If you have the money to cover the purchase, the credit card can allow you to leave the money in the bank and not carry it with you. After you make the purchase, you can pay off the balance on the credit card immediately after the purchase. Furthermore, if you pay the entire balance on a credit card each month, most credit card companies do not impose any finance charges. Above all, don't spend money you do not have.

Section 1.4 Exercises

1. Mary bought a used car for $3,500, which was financed for $160 a month for 24 months.

    a. What is the total cost for the car?

    b. How much is the finance charge?

2. Jose borrowed $800 from his local credit union for 8 months at 6% simple interest. He agreed to repay the loan by making eight equal monthly payments.

    a. How much is the finance charge?

    b. What is the total amount to be repaid?

    c. How much is the monthly payment?

3. Pam purchased a new laptop, which was advertised for $750. She bought it on the installment plan by paying $50 at the time of purchase and agreeing to pay the balance plus 18% simple interest on the balance in 24 monthly payments.

    a. How much is the finance charge?

    b. How much will each payment be?

    c. What is the total cost of the laptop?

4. Joe buys a new stereo for $875 and pays for it over two years with 6.5% simple interest. What is his monthly payment?

5. How many days are in a credit card billing cycle that runs from July 17 through August 16?

6. How many days are in a credit card billing cycle that ends on December 12?

7. How many days are in a credit card billing cycle that ends on September 22?

8. If a credit card has a balance of $30 for 11 days and then a balance of $50 for 20 days, what is the average daily balance for that 31-day period?

9. If a credit card has a balance of $120 for 10 days, a balance of $150 for the next 12 days, and a balance of $90 for 8 days, what is the average daily balance for that 30-day period?

10. If the APR on a credit card is 6.57%, what is the daily interest rate? Round your answer to the nearest thousandth of a percent.

11. If the APR on a credit card is 8.75%, what is the daily interest rate? Round your answer to the nearest thousandth of a percent.

12. The average daily balance is $210.39 on a credit card with an APR of 13.4% over a billing period of 30 days. What will the finance charge be?

13. The average daily balance is $90.15 on a credit card with an APR of 5.4% over a billing period of 30 days. What will the finance charge be?

14. The balance on Maria's credit card on May 10, the billing date, was $3,198.23. She sent in a $1,000 payment, which was posted on May 13, and made no other transactions during the billing cycle. Assuming the APR on the card is 7.9%, answer the following.

    a. What is the average daily balance for the billing period?

    b. How much is the finance charge for the billing period?

    c. What will be the new balance when she receives her June 10 statement?

15. The balance on Jim's credit card on April 17, the billing date, was $78.30. For the period ending May 16, he made the following transactions. Assume an APR of 5.2%.

    • April 20, Payment, $50.00

    • May 1, Charge: Gas, $29.20

a. Find the average daily balance for the billing period.

b. Find the finance charge that is due on May 17.

c. Find the balance that is due on May 17.

16. The balance on Carrie's credit card on November 19, the billing date, was $238.50. For the period ending December 18, she made the following transactions. Assume an APR of 2.9%.

- November 23, Payment, $150.00

- December 1, Charge: Clothing, $129.19

- December 10, Charge: Dinner, 37.43

a. Find the average daily balance for the billing period.

b. Find the finance charge that is due on December 19.

c. Find the balance that is due on December 19.

17. The balance on Ted's credit card on August 9, the billing date, was $1,298.51. For the period ending September 8, he made the following transactions. Assume an APR of 3.9%.

- August 14, Payment, $350.00

- August 22, Charge: Gasoline, $39.92

- August 29, Charge: Groceries, $87.17

a. Find the average daily balance for the billing period.

b. Find the finance charge that is due on September 9.

c. Find the balance that is due on September 9.

Answers to Section 1.4 Exercises

1. a. $3840   b. $340

2. a. $32 (use T = 8/12)   b. $832   c. $104

3. a. $252   b. $39.67 ($39.59 for the last one) c. $1002

4. $41.20 for the first 23 months, $41.15 for the last payment.

5. 31   6. 30   7. 31

8. $42.90   9. $124.00   10. 0.018%

11. 0.024%   12. $2.32   13. $0.40

14. a. $2295.00   b. $15.40   c. $2213.63

15. a. $48.87   b. $0.21   c. $57.71

16. a. $ 197.24   b. $0.47   c. $255.59

17. a. $1059.07   b. $3.51   c. $1079.11

1.5 Home Sweet Home (Mortgages)

Til Death Do Us Part

In French, the word *mort* means death and the word *gage* means pledge. Thus, the literal meaning of the word mortgage is "death pledge." Now that's something to think about …

Amortization

Amortization is a situation in which the borrower agrees to make regular payments on the principal and interest until a loan is paid off. There are a lot of variables involved in obtaining a home loan, and each should be weighed carefully before one enters into such a major commitment.

Amortization Tables, Spreadsheets, & Online Mortgage Calculators

Somewhere, someone once sat down and, hopefully using a computer, calculated the monthly payments for loans using different amounts, interest rates, and time periods. Extensive **amortization tables** are rarely seen, but they are pretty large and fairly organized. Succinct versions of these tables are a nice way to quickly compare calculations for different time periods. If you use a table computed on a fixed amount of $1,000, you can scale the amount to match any principal, being sure to round the calculated amount to the nearest cent.

Here is a partial amortization table.

Rate (%)	15 Years	20 Years	30 Years
Monthly Mortgage Payment per $1,000			
4.50	7.6499	6.3265	5.0669
4.75	7.7783	6.4622	5.2165
5.00	7.9079	6.5996	5.3682
5.25	8.0388	6.7384	5.5220
5.50	8.1708	6.8789	5.6779
5.75	8.3041	7.0208	5.8357
6.00	8.4386	7.1643	5.9955
6.25	8.5742	7.3093	6.1572
6.50	8.7111	7.4557	6.3207
6.75	8.8491	7.6036	6.4860
7.00	8.9883	7.7530	6.6530
7.25	9.1286	7.9038	6.8218
7.50	9.2701	8.0559	6.9921
7.75	9.4128	8.2095	7.1641

Using the table to find the monthly payment for a 7% loan for 30 years, we see the payment factor per $1,000 is 6.6530. Then, since the numbers in the table are given as per $1,000, to determine the monthly payment for a $120,000 loan, we multiply the 6.6530 by 120, to get a monthly payment of $798.36.

EXAMPLE 1: Use the amortization table to find the monthly mortgage payment for a $135,000 loan at 5.5% for 30 years.

Using the table, we find the payment factor of 5.5% for 30 years to be 5.6779. Then, 5.6779 × 135 give us a monthly payment of $766.5165, which has to be rounded to $766.52.

Unfortunately, amortization tables are often incomplete and even hard to locate. Alternatively, we could use a formula and a spreadsheet to determine and compare the monthly payments on a loan with different terms. To do so, we will need the following formula:

The monthly payment, M, on a loan is found by:

$$M = P\left(\frac{r/12}{1-\left(1+r/12\right)^{-12t}}\right)$$

In the formula, P indicates the original principal, r is the annual interest rate as a decimal, and t stands for the term of the loan, in years. On the surface, using the formula appears to be a bit tedious, but when it is correctly entered into a spreadsheet calculation, the computer will do all the work for us.

To generate an amortization table in a spread sheet, such as Excel® (shown below), use 1,000 for the principal, use a cell indicator for the rate (in the example below, that is the value in column A), and use the numeric value for the number of years (in the example below, that is the 15—for 15 years). Once we establish the formula, we can quickly copy it and the increasing rates for as many rows as we desire.

	f_x	=1000*(((A2/100)/12)/(1-(1+(A2/100)/12)^(-12*15)))				
	A	**B**	**C**	**D**	**E**	**F**
1	Rate	T = 15 Years				
2	4.50	7.6499				
3	4.75	7.7783				
4	5.00	7.9079				
5	5.25	8.0388				
6		—				

Select a cell and enter the monthly payment formula using a cell indicator for the rate and the # for the years.

Notice the stated rates are given as percents. Thus, in the formula appearing in the spreadsheet, the rate is changed to a decimal by dividing it by 100. Then, once we have the basic table created, we can make additional columns to cover different time periods, adjust the cells to control the number of decimal places that get displayed, and add other cosmetic effects—such as colors and cell borders.

This leads us to what can easily be considered one of the most useful websites we will ever encounter: http://www.interest.com/calculators/. Since we are only going to scratch the surface of the possibilities for the mortgage calculators on this site, you are strongly encouraged to explore it on your own. In fact, it would be wise to open that page in its own window, and bookmark (or add it to your favorites folder) it for future reference. Do realize, since the values printed in the amortization table shown in this section are rounded to four decimal places, the online calculators will be a little more accurate. But that difference, if any, should not be more than a couple of cents.

1.5 Home Sweet Home (Mortgages)

EXAMPLE 2: Use the calculators at interest.com (URL listed above) to find the monthly mortgage payment for a $135,000 loan at 5.5% for 30 years.

Go to the calculator webpage and select the Mortgage Calculator. Type in the mortgage amount, term, and rate (ignore the other information boxes for now), and then scroll down and click the calculate button to find the monthly payment is $766.52. It's that simple!

Once the monthly payment is calculated, the total cost of the home can easily be found. In the example above, the 30-year loan would involve 360 payments of $766.52, or a total cost of $275,947.20. That means the finance charge for this loan is over $140,000, which is more than the cost of the house! This is typical, and, unfortunately, just a part of the home-buying process.

EXAMPLE 3: Mason wishes to borrow $180,000, and he qualifies for a rate of 5.00%. Find the monthly payment for this loan if the term is 30 years. By how much will the payment go up if he reduces the loan to 20 years?

For 5%, the payment factor for 30 years is 5.3682, which makes the monthly payment $966.28. For 20 years, the payment factor is 6.5996, which makes the monthly payment $1,187.93. So, for the shorter term, the monthly payment is $221.65 higher.

Additionally, for the 30-year loan, the total amount to be paid for the home is $360 \times 966.28 = \$347,860.80$. Similarly, for the 20-year loan, the total amount to be paid is $285,103.20, which is a savings of $62,757.60.

The Real Monthly Payment

The monthly payment involved with a loan amount will constitute the majority of the cost of home ownership, but there are other costs that must be considered, as well. Probably the most important of all topics in this section is the calculation of the REAL monthly payment.

The complete monthly payment includes the mortgage payment, real estate taxes, and homeowner's insurance. Additionally, if the outstanding principal is more than 80% of the value of the home, private mortgage insurance (PMI) may also be required. However, since it is not required of every homeowner, to simplify our calculations a little bit, we will not deal with PMI.

EXAMPLE 4: Find the entire monthly mortgage payment for the following.

Assessed Value of the House: $120,000 |
Loan Amount: $105,000
Interest Rate: 8%
Term: 30 years
Real Estate Taxes: 2.5% of assessed value (annually)
Homeowners Insurance: $480 per year

Using the website or an amortization table from earlier in this section, the monthly loan payment for principal and interest is found to be $770.45. Taxes will cost (0.025) × ($120,000) = $3,000 per year, which adds $250 per month to the payment. Insurance, at $480 per year, will add $40 per month to the payment.

So, our total monthly payment will be $770.45 + $250 + $40 = $1,060.45.

EXAMPLE 5: Find the total monthly payment for a $175,000 mortgage loan at 6.5% for 30 years. The assessed value of the home is $200,000. The annual taxes on the home are 1.5% of the assessed value, and the insurance on the home costs $600 per year.

Using the amortization table, the payment factor is 6.3207, which makes the mortgage payment $1,106.12 per month. The taxes are $3,000 per year, which adds to $250 per month, and the monthly insurance payment is $50.

Thus, the total monthly payment is $1,406.12.

Affordability Guidelines

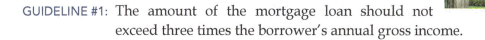

There are a few traditional guidelines that help consumers determine how much they should spend on housing costs.

GUIDELINE #1: The amount of the mortgage loan should not exceed three times the borrower's annual gross income.

Remember, this is just a consideration of the cost of the home and is not a guideline on the actual monthly payment.

GUIDELINE #2: If a family has other significant monthly debt obligations, such as car payments, credit cards, or student loans, the family's monthly housing expenses, including mortgage payment, property taxes, and private mortgage insurance, should be limited to no more than 25% of their monthly gross income (income prior to deductions).

Taking the first two guidelines into account, a family with an annual gross income of $45,000, should not purchase a home that costs more than $135,000, and the monthly expenses for the mortgage payment, property taxes, and insurance should not exceed ($45,000) × (1/12) × (0.25) = $937.50.

Realistically, however, the factor that determines whether or not a buyer can afford a home is whether or not they can obtain a loan. When banks or other financial institutions make this decision, an alternative to Guideline #2 is, often, to allow as much as 38% of the borrower's monthly income to go to all the monthly bills, which includes housing costs and other debt obligations. That leads us to the third guideline.

GUIDELINE #3: If the family has no other significant monthly debt obligations, the monthly housing costs could reach as high as 38% of their gross monthly income.

Assuming our family from above have no other significant debt obligations—such as car payments, student loans, or credit card debt—their housing costs could reach as high as 38% of their gross income, or ($45,000) × (1/12) × (0.38) = $1425.

IMPORTANT:
When following the listed Affordability Guidelines to determine the maximum monthly payment for a home loan, be sure to note whether or not other monthly debt obligations exist. If so, Guideline #2. If not, use Guideline #3.

EXAMPLE 6: John Doe has a gross monthly income of $5,125. He has a car payment and credit card debt. According to the Affordability Guidelines, how much can he afford for a total monthly housing payment?

Since the dollar amounts for the car payment and credit card debt are not given, use Guideline #2: $(0.25) \times (\$5,125) = \$1,281.25$.

Buying Points on Your Mortgage

As if taking on a mortgage payment wasn't enough, there are also fees commonly associated with acquiring a loan. After all, the bankers and loan counselors need to get paid too, and most likely, they have their own mortgage payments to make. One of these most common fees is a discount charge that lowers the interest rate, which is often referred to as buying **points**.

Let's say we plan to buy a house. If we intend to obtain a $120,000 mortgage and wish to reduce the APR, we (or the seller) can buy points. Each point typically costs 1% of the mortgage, so 2 points on a $120,000 mortgage will cost $2,400. If the seller pays these fees, the decision is easy. If we are to pay these fees, is it in our best interest to buy these points?

Unfortunately, one point does not lower the APR by one full percent. Instead, each point purchased will lower the APR by a fraction of a percent. Furthermore, this fraction varies as much as the APR itself. One day, each point purchased could lower the APR by 0.25%—that is, the going APR is 7.25%, and one point lowers it to 7%. A week later, one point may lower the APR by 0.375%, but the APR may have gone up to 7.5%.

Lowering the APR has tremendous benefits. If you are buying a home, try to get the seller to pay for as many points as you can. However, if the seller refuses to "buy down" the interest rate, many lending institutions will often try to talk you out of purchasing points. Their argument is based on the typical homeowner only spending (on average) less than 5 years in a home, thus not remaining on the property long enough to recover the cost of the points. Admittedly, this argument does have some merit. To compare the advantages and disadvantages of buying points, consider the following calculations, and assume each point will lower the APR by 0.25%.

Case 1:	Case 2:
Borrow: $100,000 For 30 Years at 7.25% Points Purchased: None	Borrow: $100,000 For 30 Years at 6.75% Points Purchased: 2 for $2,000
Monthly Payment: $682.12 Total Interest to be Paid: $145,580.03	Monthly Payment: $648.60 Total Interest to be Paid: $133,493.79

The monthly payment in Case 2 is $33.52 less than the payment in Case 1. If we don't pay any extra on the monthly payments, it will take 60 payments (60 x $33.52 = $2011.12) to recover the initial $2,000 cost of the points. If we can afford an extra $2,000 in closing costs, hope to stay in the home for the term of the loan (or at least more than 5 years), buying 2 points will save us over $12,000 in interest—making only the minimum monthly payment!

If, however, we plan to remain in the home for less than 5 years, intend to make only the minimum payments, and wish to keep the closing costs as low as possible, buying points is a bad idea. If we are buying a "starter" home and plan to move to a bigger home in a few years, it is a good idea to pay as little as possible and save our money for the future investment.

EXAMPLE 7: Nathan wishes to borrow $225,000 to purchase a home. He will get a 30-Year mortgage, and the bank is offering a rate of 6.25%. If he buys down the rate to 5.75%, by how much will he reduce his monthly payment? If he pays $4,500 for the points, how long will it take for him to recover the cost?

For the 6.25% rate, the monthly payment is $1,385.36. For the 5.75% rate, the payment is $1,313.04, which is a savings of $72.32 each month.

$4,500 divided by $72.32 per month, tells us it will take 62.2 months (or 63 *whole* months) to recover the cost of the points.

Section 1.5 Exercises

1. Use the amortization table earlier in this section to find the monthly mortgage payment for loans with the following conditions.

    a. $150,000 for 20 years at 5.25%

    b. $172,000 for 30 years at 7.00%

    c. $210,000 for 30 years at 5.75%

    d. $275,000 for 15 years at 5.50%

2. Use the online mortgage calculators found at http://www.interest.com/calculators/ to find the monthly mortgage payment for loans with the following conditions.

    a. $150,000 for 20 years at 5.25%

    b. $172,000 for 30 years at 7.00%

    c. $210,000 for 30 years at 5.75%

    d. $275,000 for 15 years at 5.50%

3. Dave wishes to borrow $210,000, and he qualifies for a rate of 7.25%. Find the monthly payment for this loan if the term is 30 years. By how much will the payment change if he reduces the loan to 20 years?

4. Trevor wishes to borrow $280,000, and he qualifies for a rate of 5.50%. Find the monthly payment for this loan if the term is 30 years. By how much will the payment change if he reduces the loan to 15 years?

5. Laura wishes to borrow $175,000, and she qualifies for a rate of 6.25%. Find the monthly payment for this loan if the term is 20 years. By how much will the payment change if she takes out the loan to 30 years?

6. Find the total monthly payment for a $275,000 mortgage loan at 5.5% for 30 years. The assessed value of the home is $300,000. The annual taxes on the home are 1.2% of the assessed value, and the insurance on the home costs $750 per year.

7. Find the total monthly payment for a $205,000 mortgage loan at 5.25% for 20 years. The assessed value of the home is $240,000. The annual taxes on the home are 1.25% of the assessed value, and the insurance on the home costs $450 per year.

8. Find the total monthly payment for a $185,000 mortgage loan at 4.5% for 30 years. The assessed value of the home is $200,000. The annual taxes on the home are 0.9% of the assessed value, and the insurance on the home costs $375 per year.

9. According to the Affordability Guidelines, if a family has an annual income of $62,000, what is the maximum mortgage amount the family could afford?

10. According to the Affordability Guidelines, if a family has a gross annual income of $122,000, what is the maximum mortgage amount the family could afford?

11. According to the Affordability Guidelines, if a family has a gross annual income of $312,000, what is the maximum mortgage amount the family could afford?

12. According to the Affordability Guidelines, if a family has a gross annual income of $63,000 and other significant debt obligations, what should be the maximum monthly mortgage payment the family could afford?

13. According to the Affordability Guidelines, if a family has a gross annual income of $174,000, and other significant debt obligations, what should be the maximum monthly mortgage payment the family could afford?

14. According to the Affordability Guidelines, if a family has a gross annual income of $48,000 and other significant debt obligations, what should be the maximum monthly mortgage payment the family could afford?

15. According to the Affordability Guidelines, if a family has a gross annual income of $33,000, and no other significant debt obligations, what could be the maximum monthly mortgage payment the family could afford?

16. According to the Affordability Guidelines, if a family has a gross annual income of $120,000, and no other significant debt obligations, what could be the maximum monthly mortgage payment the family could afford?

17. According to the Affordability Guidelines, if a family has a gross annual income of $57,000, and no other significant debt obligations, what could be the maximum monthly mortgage payment the family could afford?

18. How much will two points cost on a $175,000 mortgage?

19. How much will three points cost on a $275,000 mortgage?

20. How much will two points cost on a $215,000 mortgage?

21. How much will one point cost on a $185,000 mortgage?

22. Monica wishes to borrow $325,000 to purchase a home. She will get a 30-Year mortgage, and the bank is offering a rate of 5.75%. If she buys down the rate to 5.25%, by how much will she reduce her monthly payment? If she pays $6,500 for the points, how long will it take for her to recover the cost?

23. Juan wishes to borrow $185,000 to purchase a condo. He will get a 20-Year mortgage, and the bank is offering a rate of 5.50%. If he buys down the rate to 5.25%, how much will he reduce his monthly payment by? If he pays $1,850 for the single point, how long will it take for him to recover the cost?

Answers to Section 1.5 Exercises

1. a. $1010.76   b. $1144.32   c. $1225.50   d. $2246.97

2. a. $1010.77   b. $1144.32   c. $1225.50   d. $2246.98

3. The payment will go up $227.22.

4. The payment will go up $698.02 (using interest.com) or $689.01 (using the tables).

5. The payment will go down $201.61 (using interest.com) or $201.62 (using the tables).

6. $1923.92

7. $1668.88 (using interest.com) or $1668.87 (using the tables)

8. 8.$1,118.62 (using interest.com) or $1,118.63 (using the tables)

9. 9.$186,000

10. $366,000   11. $936,000   12. $1312.50   13. $3625.00

14. $1000.00   15. $1045.00   16. $3800.00   17. $1805.00

18. $3500.00   19. $8250.00   20. $4300.00   21. $1850.00

22. The monthly payment will go down $101.95. It will take 64 months (5 years, 4 months) to recover the cost of the points.

23. The monthly payment will go down $25.98 ($25.99, if using the tables.). It will take 71.15 months (72 *whole* months, which is 6 years) to recover the cost of the point.

*NOTE: When no method was specified and if you used an amortization table, the monthly payment formula, or an online calculator, it is possible your answers may have been off by a cent or two. For the purposes of the material presented in this section, that is acceptable. If your answers were off by a greater amount, you may have either made a calculation mistake or inappropriately rounded a value at an intermediate step.

Credits

1. "Money Stacks," http://pixabay.com/en/dollar-money-finance-dollars-163473/. Copyright in the Public Domain.

2. Copyright © Wikimedia Foundation, Inc. (CC BY-SA 3.0) at http://en.wikipedia.org/wiki/Benjamin_Franklin.

3. "Benjamin Franklin," http://commons.wikimedia.org/wiki/File:Benjamin_Franklin.PNG. Copyright in the Public Domain.

4. Copyright © Wikimedia Foundation, Inc. (CC BY-SA 3.0) at http://en.wikipedia.org/wiki/Dow_Jones_Industrial_Average.

5. Copyright © Wikimedia Foundation, Inc. (CC BY-SA 3.0) at http://en.wikipedia.org/wiki/Ponzi_scheme.

6. Copyright © Wikimedia Foundation, Inc. (CC BY-SA 3.0) at http://en.wikipedia.org/wiki/Charles_Ponzi.

7. "Charles Ponzi," http://commons.wikimedia.org/wiki/File:Charles_Ponzi.jpg. Copyright in the Public Domain.

8. Copyright © Blane (CC by 3.0) at http://commons.wikimedia.org/wiki/File:Delay_lhr.jpg.

9. "Hiking Shoes," pixabay.com/en/shoes-footwear-hiking-shoes-walking-584850/. Copyright in the Public Domain.

10. "Capitol Building Full View," http://commons.wikimedia.org/wiki/File:Capitol_Building_Full_View.jpg. Copyright in the Public Domain.

11. "Froot-Loops-Cereal-Bowl," http://commons.wikimedia.org/wiki/File:Froot-Loops-Cereal-Bowl.jpg. Copyright in the Public Domain.

12. "Computer Server," http://pixabay.com/en/computer-server-workstation-hosting-158474/. Copyright in the Public Domain.

13. "Red Books," http://pixabay.com/en/books-pile-red-reader-reading-158813/. Copyright in the Public Domain.

14. "Coffee Cup," http://pixabay.com/en/coffee-cup-smoking-hot-drink-34251/. Copyright in the Public Domain.

15. "Cell Phone," http://pixabay.com/en/cell-phone-mobile-technology-33083/. Copyright in the Public Domain.

16. "CD," http://pixabay.com/en/cd-music-audio-notes-mp3-sound-158817/. Copyright in the Public Domain.

17. "Chess Pieces," http://pixabay.com/en/chess-strategy-chess-board-316658/. Copyright in the Public Domain.

18. "Sunflowers," http://pixabay.com/en/arrangement-bloom-blossom-bouquet-16858/. Copyright in the Public Domain.

19. Copyright © Lotus Head (CC by 3.0) at http://commons.wikimedia.org/wiki/File:Credit-cards.jpg.

20. "Woman with Credit Card," http://pixabay.com/en/business-card-credit-debit-female-15721/. Copyright in the Public Domain.

21. "German House," http://pixabay.com/en/house-charming-mu%CC%88nster-germany-245409/. Copyright in the Public Domain.

22. "Brick Home," http://pixabay.com/en/brick-home-blue-sky-architecture-290334/. Copyright in the Public Domain.

23. "Cabin," http://pixabay.com/en/slave-cabin-laura-plantation-440349/. Copyright in the Public Domain.

24. renjith krishnan, "Concept Of Loan. House And Mousetrap Stock Photo," http://www.freedigitalphotos.net/images/concept-of-loan-house-and-mousetrap-photo-p232048. Copyright © 2014 by FreeDigitalPhotos.net. Reprinted with permission.

25. "Keys," http://pixabay.com/en/keys-key-lock-door-symbols-access-20290/. Copyright in the Public Domain.

2

SETS & VENN DIAGRAMS

People love to group things together. We put all our socks in a drawer, our books on a shelf, and our keys on a keychain. Any collection of distinct objects is called a set, and believe it or not, sets are one of the most fundamental concepts in mathematics.

Set theory, including the study of Venn diagrams, was developed around the end of the 19th century. Basic set theory concepts are often taught to children; while more advanced concepts are usually parts of a college education. Despite the simplicity of merely placing objects together, college-level set theory can be quite rigorous.

2.1 On the Shoulders of Giants (Biographies & Historical References)

Much of what we study in relation to sets and Venn diagrams can be traced back to two individuals who provided many of their contributions around the end of the 19th century. Although there is no evidence suggesting the Englishman John Venn and the Russian-born Georg Cantor worked in collaboration, they were likely aware of each other's work and probably shared correspondences with other mathematicians from the same era.

John Venn

John Venn (1834–1923) was born in Hull, England. His mother, Martha Sykes, died when he was just three years old, and his father was the Reverend Henry Venn. John was descended from a long line of church evangelicals, including his grandfather John Venn, who led a sect that campaigned for the abolition of slavery, advocated prison reform and the prevention of cruel sports, and supported missionary work abroad. He studied with his brother, Henry, in London in 1846, and moved on to Cambridge in 1853. In 1857, he obtained his degree in mathematics and would eventually follow his family vocation and become an Anglican priest, ordained in 1859.

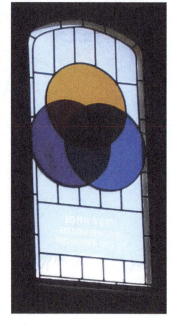

Venn returned to Cambridge in 1862, as a lecturer in moral science, studying and teaching logic and probability theory, and, beginning around 1869, giving intercollegiate lectures. These duties led to his developing the diagram, which would eventually bear his name. A stained glass window in one of the dining halls at the university, shown here, commemorates Venn's work.

He resigned from the clergy in 1883, having concluded that Anglicanism was incompatible with his philosophical beliefs. In that same year, Venn was elected a Fellow of the Royal Society and was awarded a Doctor of Science by Cambridge.

In 1903 he was elected President of the Gonville and Caius College at Cambridge, a post he held until his death, on April 4, 1923. In commemoration of the 180th anniversary of Venn's birth, on August 4, 2014, Google replaced its normal logo on global search pages with an interactive and animated "Google Doodle" that incorporated the use of a Venn diagram.[1,2]

Georg Cantor

Georg Ferdinand Ludwig Philipp Cantor (1845–1918) was born in the western merchant colony in Saint Petersburg, Russia, and brought up in the city until he was eleven. Georg, the oldest of six children, was regarded as an outstanding violinist. Seeking winters milder than those of Saint Petersburg, his family moved to Germany in 1856. With exceptional skills in mathematics, in 1862, Cantor entered the University of Zürich, and later studied at the University of Berlin, as well as the University of Göttingen. In 1867, Cantor completed his dissertation, on number theory, at the University of Berlin.

Cantor's work between 1874 and 1884 is the origin of set theory. Prior to this work, the concept of a set was a rather elementary one. Before Cantor, there were only finite sets, which are relatively easy to understand. Everything else was "the infinite" and was considered a topic for philosophical, rather than mathematical, discussion. By proving that there are infinitely many possible sizes for infinite sets, Cantor established that set theory was not trivial, and it needed to be studied further. The basic concepts of set theory brought to the forefront by Cantor are now used throughout mathematics.

Cantor defined infinite and well-ordered sets, and proved that the real numbers are "more numerous" than the natural numbers. As much of his work centered on the concept of infinity, many of his theories were originally regarded as so counter-intuitive—even shocking—that they encountered resistance from mathematical contemporaries such as Leopold Kronecker

---

1   Copyright © Wikimedia Foundation, Inc. (CC BY-SA 3.0) at http://en.wikipedia.org/wiki/John_Venn.
2   Copyright © New World Encyclopedia (CC BY-SA 3.0) at http://www.newworldencyclopedia.org/entry/ John_Venn.

and Henri Poincaré. In a time where many Christians believed everything was tied to divinity, Kronecker's objections to Cantor's work on set theory were so vehement, he was even quoted as having said, "God made the integers, all else is the work of man."

The objections to Cantor's work were occasionally fierce: Poincaré referred to his ideas as a "grave disease" infecting the discipline of mathematics, and Kronecker's public opposition and personal attacks included describing Cantor as a "scientific charlatan", a "renegade" and a "corrupter of youth" for teaching his ideas to a younger generation of mathematicians. Cantor's recurring bouts of depression from 1884 to the end of his life have been blamed on the hostile attitude of many of his contemporaries. In the 1880s heavy criticism of his work took a toll on Cantor's self-confidence, as he shifted his activities toward philosophy, and away from mathematics. He also began an intense study of Elizabethan literature, thinking there might be evidence that Francis Bacon wrote the plays attributed to William Shakespeare.

Cantor retired in 1913, living in poverty and suffering from malnourishment during World War I. The public celebration of his 70th birthday was canceled because of the war, and he died on January 6, 1918, in the sanatorium where he had spent the final year of his life.[3]

2.2 On Your Mark. Get Set. Go! (Basic Set Concepts & Notations)

Forms of Sets

In order to better organize data, we often put them into sets. Simply stated, a **set** is a collection of distinct objects. Sets can be indicated several different ways, and three of the most common ways are using **roster form**, **set-builder notation**, or **verbal** (or written) **description**. **NOTE:** In set-builder notation, the symbol | is translated as "such that."

Notation	Example
Roster form	{2, 3, 5, 7, 11, 13, 17, 19}
Set Builder notation	{x | x is a prime number less than 20}
Verbal Description	The set of all prime numbers less than twenty.

One method to indicate a set is not necessarily better than another; they're just different.

If it is not possible (or practical) to list all of a set's elements using the roster notation, an **ellipsis** (" ... ") can be used. But, before an ellipsis can be used, enough elements must be listed in order to establish a pattern. For example: {2,..., 19} could mean all the prime numbers less than 20 or, possibly, all of the whole numbers from 2 to 19. When in doubt, it is usually best to describe larger sets with set-builder notation or a verbal description.

When writing a set in roster form, we need to be sure to separate each element with a comma. Also, be sure to use braces and not parentheses! For example, to indicate a set consisting of the lowercase letters a, b, and c in roster form, we must type the set as **{a, b, c}**. {a b c} and (a, b, c) are both incorrect because {a b c} lacks commas, and (a, b, c) uses parentheses.

> IMPORTANT: When writing ANY set in roster form or set-builder notation, we must be sure to encase the elements within a pair of braces, { }. The use of any other grouping symbols (parentheses or brackets) is incorrect

When is a Letter Not a Letter?

In math, the symbols we use are just that—symbols. And, to make life easier, we frequently utilize the characters at our fingertips, which, more often than not, are the letters appearing on a computer keyboard. Keep in mind, however, they are just symbols. That means, since "A" and "a" are different characters, they are also different symbols. Even though they are both versions of the same letter, when we use them as symbols, they are as different as "A" and "B."

Think of the characters Δ and δ. Would you call them the same? Δ is the capital Greek letter delta, and δ is how we write the lowercase delta. In math, we would not think of them as Greek letters; we use them as symbols. The letters in the English alphabet need to be treated the same way.

Taking into account we are working with symbols and not letters …

"A" is not an element in the set {a, b, c}, but "a" *is* an element of the set {a, b, c}. Using the symbols for "element" and "not an element," we have:

$A \notin \{a, b, c\}$ and $a \in \{a, b, c\}$

It is, however, worth noting that it is customary to use lowercase English letters as the symbols for elements and capital English letters as symbols for sets. We will see more of this in the next section.

Another subtle but important characteristic of a set involves the word "distinct." To be precise, when listing a set, we should not duplicate the elements in the list. For example, {a, b, b, c} is a list of objects, but we should not call it a set, because the b is listed twice. The order of the elements does not matter, but they should not be duplicated. Since we are doing a relatively short exploration into the topic of Set Theory, this rule doesn't really matter for our purposes. Thus, we will abuse this rule from time to time and refer to lists like {a, b, b, c} as sets. Please afford us this luxury.

Infinite & Finite Sets

An **infinite set** has an unlimited number of elements. For example, the set of whole numbers, {0, 1, 2, 3, … } is an infinite set. If we know how many elements are in the set, or we know there is an end to the number of elements, we have a **finite set**. For example, the set {a, b, c} is a finite set that contains three elements.

EXAMPLE 1: Are the following sets infinite or finite?

    a. The set of capital letters in the English alphabet Finite

    b. The set of whole numbers Infinite set

    c. All the gold bars in Fort Knox Finite

Answers

    a. Finite. There are 26 of them.

    b. Infinite. If you think you have the last one, just add 1 to it.

    c. Finite. We may not know how much is there, but there is a specific amount.

Well Defined Sets

If there is no ambiguity or subjectivity as to whether an element belongs to a set, the set is known as **well defined**. In other words, the set of integers is well defined, since it is clear whether or not a number is in that set. The set of good teachers at a school is not well defined, because it is not clear whether Teacher X is to be considered "good" or not.

EXAMPLE 2: Determine whether the following sets are well defined.

    a. The set of capital letters in the English alphabet. Fi

    b. The set of keys on John's keychain.

    c. The 10 best Disney characters of all time.

Answers

    a. Since we know the exact letters in the alphabet, the set is well defined.

    b. Although we may not know what keys are on the keychain, it is possible to determine exactly which keys are there. Also, no matter who makes the determination, the results are exactly the same every time. Thus, the set is well defined.

c. Since two different people will make two different lists, the set is not well defined. Yes, we know there are 10 characters in the list, but we do not know which 10 characters are there.

Notations & Symbols

Symbol	Meaning	Example
~	Equivalent	{a, b, c} ~ {#, $, %} The sets have same number of elements.
=	Equal	{a, b, c} = {c, b, a} The sets contain the exact same elements.
∈	Element Of	a ∈ {a, b, c}
∉	Not an Element Of	D ∉ {a, b, c}
∅ or { }	The Empty Set	The set of all U.S. states sharing a land border with Hawaii is the empty set.

Take a closer look at the symbols used for the empty set. In particular, look at the ∅. That symbol is *not* the number zero. Although many people like to distinguish between the letter O and the number 0 by placing a diagonal slash through the number 0, the ∅ symbol means something entirely different. Remember, the empty set has no elements in it. {0} is the set containing the number 0, and { } (or ∅) is the empty set.

Think of it this way: There is a big difference between having a checking account with no money in it and not having a checking account at all. For the former, you can walk into your bank, ask for your account balance, and be told, "Your balance is 0." If, however, you do not have an account (or you go into the wrong bank), the teller cannot quote you a balance for an account that does not exist.

If you are in the habit of putting slashes through your zeros, make an effort to limit your slash to the inside of the 0, itself. Then, for the empty set, extend your slash through the symbol on both the top and bottom.

0

Alternative form for the number zero.

Symbol for the empty set.

The / is completely within the zero.

The / extends through the symbol.

To help keep things straight, throughout the remainder of this text, we will see the symbols for the empty set—both { } and ∅—interchangeably, and we will never put a / through any of our zeros.

EXAMPLE 3: Each of the following statements is false. Without negating the symbol, rewrite the statement, so it is true. By the way, there are several ways to "fix" each statement.

a. $7 \in \{2, 4, 6, 8, 10\}$

c. $\{d, f, g\} \sim \{hat, coat, gloves, scarf\}$

b. $\{1, 2, 3, 4, 5, 6\} = \{q, w, e, r, t, y\}$

d. $\{a, b, c\} = (c, a, b)$

Possible Corrections

a. Change the 7 into one of the elements in the set, or put 7 into the set.

b. Change the = to ~.

c. Add an element to the set on the left, or remove one from the set on the right.

d. Change the parentheses to braces. Without braces, the list on the right is not even a set.

Section 2.2 Exercises

For Exercises #1 through #4, rewrite the statement using set notation.

1. 5 is not an element of the empty set.

2. t is an element of set B.

3. $13 < x \leq 19$

4. $x = 31$

For Exercises #5 through #9, list the elements in roster form.

5. The two-digit numbers on a standard analog clock.

6. The months of the year that do not have 31 days.

7. The set of all the states that touch the Pacific Ocean.

8. The set of all the oceans that touch Nebraska.

9. {x | x is an odd whole number less than 10}

For Exercises #10 through #20, identify the statements as true or false.

10. Two sets that are equal must also be equivalent.

11. Zero is an element of the empty set.

12. The set of good students in our class is a well-defined set.

13. {0} is an empty set

14. {s, p, a, m} = {m, a, p, s}

15. {q, u, i, c, k} is a finite set.

16. Every well-defined set must also be finite.

17. The set of purple jellybeans is well defined.

18. The set of purple jellybeans is infinite.

19. If two sets are equivalent, then they must be equal.

20. The set of ugly dachshunds is well defined.

For #21 through #29, each of the statements is false. Provide at least two possible corrections. In each correction, change only one thing in the statement.

21. 14 ∈ {2, 4, 6, 8, 10, 12}

22. {Fred, Barney, Betty, Wilma} ~ {Dino, Pebbles, Bam Bam}

23. $A \in \{a, b, c, d, f\}$

24. hat $\in \{$shirt, pants, shoes, coat, socks$\}$

25. $\{1, 3, 4\} = \{a, c, e\}$

26. The set of people over 6 feet tall is an infinite set.

27. $a \notin \{a, e, i, o, u\}$

28. $\{1, 2, 3, 4, 5, 6, 7, \ldots\}$ is not well defined.

29. $\{h, i\} = \{hi\}$

30. What is the empty set?

Answers to Section 2.2 Exercises

1. $5 \notin \{\}$ or $5 \notin \varnothing$

2. $t \in B$

3. $\{x \mid 13 < x < 19\}$

4. $\{31\}$

5. $\{10, 11, 12\}$

6. {February, April, June, September, November}

7. {Alaska, Washington, Oregon, California, Hawaii}

8. $\{\}$

9. $\{1, 3, 5, 7, 9\}$

10. True

11. False

12. False

13. False

14. True

15. True

16. False

17. True

18. False

19. False

20. False

For Exercises #21 through #29, answers may vary.

21. $14 \notin \{2, 4, 6, 8, 10, 12\}$, or
    $12 \in \{2, 4, 6, 8, 10, 12\}$

22. {Fred, Barney, Betty} ~ {Dino, Pebbles, Bam Bam}, or
    {Fred, Barney, Betty, Wilma} ~ {Dino, Pebbles, Bam Bam, Stan}

23. $A \in \{a, b, c, d, f\}$, or $a \in \{a, b, c, d, f\}$

24. hat $\notin$ {shirt, pants, shoes, coat, socks}, or
    hat $\in$ {shirt, pants, shoes, coat, socks, hat}

25. $\{1, 3, 4\}$ ~ $\{a, c, e\}$, or
    $\{1, 3, 4\} \neq \{a, c, e\}$

26. The set of people over six feet tall is not an infinite set., or The set of people over six feet tall is a finite set.

27. $a \in \{a, e, i, o, u\}$, or
    $y \notin \{a, e, i, o, u\}$

28. $\{1, 2, 3, 4, 5, 6, 7, \ldots\}$ is well-defined., or
    $\{1, 2, 3, 4, 5, 6, 7, \ldots\}$ is infinite.

29. $\{h, i\} = \{h, i\}$, or $\{h, i\} \neq \{hi\}$

30. A set that contains no elements.

2.3 How Many are There? (Subsets & Cardinality)

Subsets

When all of the elements of one set are found in a second set, the first set is called a **subset** of the second set. A formal definition is:

If every element in set B is also an element in set A, then B is a subset of A.

EXAMPLE 1: Given set A = {a, b, c, d, e}, B = {a, e}, and C = {c, d, f}

    a. Since all of the elements in B are also in set A, set B is a subset of A.

    b. Set C is not a subset of A, because the letter "f" is not a member of set A.

Every set has a certain number of subsets, and sometimes, it is useful to find all of them. We can determine the number of subsets through the use of a systematic listing process.

EXAMPLE 2: Let's say we would like to order a pizza from Phony Pizza, and the only toppings they offer are pepperoni, mushrooms, and anchovies. Taking into consideration that cheese has to be on every pizza Phony Pizza makes, let's determine all the possible pizzas we could order.

Here is a set that contains all of the possible toppings: {p, m, a}. What we have been asked to do here is to find all of the subsets of that set.

The one-topping pizzas we could order are: {p}, {m}, and {a}

The two-topping pizzas we could order are: {p, m}, {p, a}, and {m, a}

We could order a pizza with all the toppings: {p, m, a}

So far we have found seven different pizzas we could order, but there is one more that we don't want to forget. We could also order a pizza with no toppings—a plain cheese pizza. Since this pizza would contain none of the toppings, we would include this pizza in our list as the **empty set** and use either { } or ∅ to represent it. Thus, in total, there are eight different pizzas that we could order, which means the set {p, m, a} has a total of eight subsets. Sometimes it is difficult to accept the empty set as a subset. Look back at the "digestible" example, and ask yourself the following question: Is there any element of { } that is not contained in {p, m, a}?

Remember, there are no elements in the empty set. Thus, in the terms of our pizza toppings, there does not exist a topping in { } that is not in the set {p, m, a}. In other words, in order to *not* be a subset, there would have to be a topping listed in { } that you could not find in {p, m, a}. Given that there are no toppings in { }, there is nothing to compare with {p, m, a}. Since the empty set qualifies as a subset, this leads us to following important fact.

> By definition, the empty set is a subset of every set.

Also, please note the entire set also is a subset of itself. That is, {p, m, a} is, indeed, a subset of {p, m, a}.

Let's look at some other examples of subsets that don't involve pizza.

A Given Set	{a}	{a, b}	{a, b, c}
Subsets of the set containing zero elements	{ }	{ }	{ }
Subsets of the set with one element	{a}	{a}, {b}	{a}, {b}, {c}
Subsets of the set with two elements	–	{a, b}	{a, b}, {a, c}, {b, c}
Subsets of the set with three elements	–	–	{a, b, c}
Total number of subsets for the given set	2	4	8

If we continue to look at and count the number of subsets for a given set, we would notice an important fact. The total number of subsets of a given set—remember to include the empty set and the whole set, itself—is 2^n, where n is the number of elements in the given set.

In the table listed above, the set {a, b, c} has three elements. Thus, there are $2^3 = 8$ subsets for it.

Intersection, Union, & Complement

Formally, the **intersection** of set A and set B, denoted A ∩ B, is the set that consists of all the elements in *both* sets. That is, in order for an element to be in the intersection of set A and set B, that element must be in set A *and* in set B. Simply put, the intersection of two sets is the set of elements they have in common, or where the two sets overlap.

The Word "Or"

When we go to a restaurant and are presented with the choice of "soup or salad," this generally means we get to choose one item or the other, but not both. Here, the restaurant is using **exclusive "or,"** meaning the option of choosing both things has been excluded.

It is certainly possible, however, to use the word "or" in such a way that does not eliminate the possibility of having both. Let's say a friend asks you if you want to have "ketchup or mustard" on your hot dog. In this case, you could have ketchup only, mustard only, or you could have both. Here, your friend is using **inclusive "or,"** meaning the option of having both things is a possibility. Unless stated otherwise, when working with basic set theory, always use inclusive "or."

The **union** of set A and set B, denoted A ∪ B, is the set that consists of all the elements in at least one of the two sets. That is, if an element is in set A or in set B, then it is in the union of the two sets. Remember, we are using inclusive "or," so this means the set A ∪ B contains all of the elements that are in set A, joined with all the elements of set B, and it does include the elements that are common to both sets—but we don't list them twice.

Often a **universal set** is given to provide a point of reference for all of the possible elements under consideration. In other words, when forming a set, we cannot include any element that is not in the universal set. The universal set will be denoted with a capital letter U. Be careful not to confuse this with the symbol for union, ∪. If set A is contained in the universal set, U, the **complement** of set A, denoted A', is the set consisting of all the elements in U that are not in A.

Notations

Symbol	Meaning	Example
⊂	Subset of	{a, b} ⊂ {a, b, c, d}
∪	Union	A ∪ B consists of all the elements in *at least one* of the two sets.
∩	Intersection	A ∩ B consists of all the elements in *both* of the sets.
'	Complement*	A' consists of all the elements in the given universal set that are *not* in set A.

*The compliment of a set can be indicated in a number of different ways, including the apostrophe (as above), a super-scripted c, or an overscore. Symbolically, these notations look like A', A^c, and $\overline{A}$. We will stick with the apostrophe notation, because, quite frankly, the other two are hard to type.

EXAMPLE 3: Given the following sets: U = {1, 2, 3, 4, 5} ; A = {1, 4, 5} ; B = {2, 4}

A ∩ B = {4} and A ∪ B = {1, 2, 4, 5}

A' = {2, 3} and B' = {1, 3, 5}

The subset symbol, ⊂ is actually for **proper subsets**. Proper subsets are subsets that are not equal to the entire parent set. The word "proper" can be bit misleading. We are not calling them right or wrong. In this case, we just mean a proper subset is smaller than the whole set. Remember our pizza example? {p, m, a} was the pizza that had all of the toppings Phony Pizza offered. It is certainly acceptable to order a pizza with everything. Just like the symbol ≤ means "less than *or* equal to," putting an underscore beneath the proper subset symbol, ⊆, gives us the "equal to" option.

By the way, the use of the word "proper" in this sense is not unique to just sets. We have seen it before. In terms of positive numbers, a proper fraction is a fraction that is less than 1, and an improper fraction is a fraction that is greater than 1. There is nothing "wrong" with an improper fraction.

That said, to be technical, {a, b, c} ⊂ {a, b, c} is incorrect, because we do not have the "equal to" option on the symbol. In order to indicate that the entire parent set is a subset, we would have to write {a, b, c} ⊆ {a, b, c}. Conversely, if we use the ⊆ symbol to indicate a proper subset, we are fine. That is, there is nothing wrong with {a, b} ⊆ {a, b, c}. Remember, ⊆ means the first set is either a proper subset OR equal to the set on the right.

EXAMPLE 4: Each of the following statements is false. Without negating the symbol, rewrite the statement so that it is true. Please note, there are several ways to "fix" each statement.

    a. $\{a, b\} \subset \{a, c, e, f\}$

    b. $\{\#, \$, \%, \&\} \cup \{\$, \#\} = \{\$, \#\}$

    c. $\{\#, \$, \%, \&\} \cap \{\$, \%\} = \{\$\}$

    d. $U = \{1, 2, 3, 4, 5\}$ and $A = \{2, 3, 4\}$, so $A' = \{1, 4\}$

Possible Corrections (Your answers may vary)

    a. Change the b to one of the elements in the set on the right (other than a), or put the b in the set on the right.

    b. Change the union to an intersection, or put the % and & in the set on the right.

    c. Put the % in the set on the right.

    d. In A', change the 4 to 5.

Remember the arithmetic order of operations? Specifically, we need to recall how parentheses are handled. In the expression $22 - 14 + 5$, we simplify it going from left to right to get 13. If, however, we have $22 - (14 + 5)$, we need to perform the operation within the parentheses first. $22 - (14 + 5) = 3$, not 13.

Unions, intersections, and complements can be thought of as the arithmetic operations for sets. Thus, if we introduce parentheses, we will need to begin any simplification by performing the operation(s) within the parentheses.

EXAMPLE 5: Given the sets U = {2, 3, 4, 5, 6, 7, 8} ; A = {5, 7, 8} ; B = {2, 4, 5, 6, 7}, find:

    a. $A \cup B$

    b. $A \cap B$

    c. A'

    d. B'

    e. $(A \cup B)'$

    f. $(A \cap B)'$

    g. $A' \cup B'$

    h. $A' \cap B'$

Answers

    a. {2, 4, 5, 6, 7, 8}

    b. {5, 7}

    c. {2, 3, 4, 6}

    d. {3, 8}

    e. {3}

    f. {2, 3, 4, 6, 8}

    g. {2, 3, 4, 6, 8}

    h. {3}

Cardinality

The term **cardinality** refers to the number of elements in a set. The notation used to denote the cardinality of set A is **n(A)** and is read as "the number of elements in set A." Cardinality is also considered with intersection and union of sets, for example: n(A ∪ B), which would be "the number of elements in the union of sets A and B."

For a finite set, cardinality of the set should be determined and given as a single number. For example, if set A is the set consisting of all capital letters in the English alphabet, state n(A) = 26.

EXAMPLE 6: For A = {a, b, c, d, e, f, g, h, i, j}, B = {a, b, c, d, e, f}, C = {a, e, i}, D = {b, c, d}, find:

  a. n(A)

  b. n(A ∪ B)

  c. n(A ∩ B)

  d. n(B ∩ C)

  e. n(C ∩ D)

Answers

  a. 10

  b. 10

  c. 6

  d. 2

  e. 0

Section 2.3 Exercises

For Exercises #1 through #5, identify the statements as true or false:

1. $\{a, f\} \subset \{a, b, c, d, e, f, g, h\}$

2. $\{1, 4, 7\} \not\subset \{1, 2, 3, 4, 5, 6, 7, 8, 9\}$

3. $\varnothing \subset \{red, white, blue\}$

4. $\{hawk, eagle, falcon, seagull\} \subseteq \{hawk, eagle, falcon, seagull\}$

5. $(a, b, c) = \{c, a, b\}$

For Exercises #6 through #8, list all the possible subsets of the given sets:

6. $\{j, r\}$

7. $\{Moe, Larry, Curly\}$

8. {m, a, t, h}

9. A set contains 5 elements. How many subsets does this set have?

10. Explain, in your own words, why the empty set is a subset of every set.

For Exercises #11 through #13, given the groups of elements, name an appropriate universal set to which the elements could belong.

11. Apple, Dell, Gateway, Hewlett Packard

12. apple, banana, kiwi, orange, strawberry

13. basset hound, beagle, dachshund, poodle, St. Bernard

For Exercises #14 through #17, use the universal set, U = {1, 2, 3, 4, 5, 6, 7, 8, 9}

14. Find {1, 3, 5}′

15. Find {2, 4, 6, 8}′

16. Find {1, 2, 3, 4, 5}′

17. Find {1, 2, 3, 4, 5, 6, 7, 8, 9}′

18. What does the notation, n(A) indicate?

For Exercises #19 through #21, U = {a, b, c, d, e, f, g}, A = {a, b, e, f}, B = {b, d, e, f, g} and C = {b, d, e}.

19. Find n(A)

20. Find n(B′)

21. Find n(C)

For Exercises #22–33, U = {0, 1, 2, 3, 4, 5, 6, 7, 8}, A = {0, 3, 4, 5}, B = {0, 2, 6}, and C = {1, 3, 5, 7, 8}.

22. Find A ∩ B

23. Find A ∪ C

24. Find B ∩ C

25. Find $B \cup C$

26. Find $A' \cup B$

27. Find $B \cup C'$

28. Find $A' \cap C'$

29. Find $(A \cap B) \cup C$

30. Find $(B \cup C)' \cap A$

31. Find $C \cap (A' \cup B)$

32. Find $n(A \cup B')$

33. Find $n(B' \cap C)$

34. Do we use inclusive "or" or exclusive "or" in this course?

35. How do you find the intersection of two sets?

36. How do you find the union of two sets?

Answers to Section 2.3 Exercises

1. True

2. False

3. True

4. True

5. False — (a, b, c) is not a set

6. ∅, {j}, {r}, {j, r}

7. ∅, {Moe}, {Larry}, {Curly}, {Moe, Larry}, {Moe, Curly}, {Larry, Curly}, {Moe, Larry, Curly}

8. ∅, {m}, {a}, {t}, {h}, {m, a}, {m, t}, {m, h}, {a, t}, {a, h}, {t, h}, {m, a, t}, {m, a, h}, {m, t, h}, {a, t, h}, {m, a, t, h}

9. 32

10. answers may vary

11. computers

12. kinds of fruit

13. dog breeds

14. {2, 4, 6, 7, 8, 9}

15. {1, 3, 5, 7, 9}

16. {6, 7, 8, 9}

17. ∅ or { }

18. This notation indicates the cardinality of set A, which is the number of elements in that set.

19. 4

20. 2

21. 3

22. {0}

23. {0, 1, 3, 4, 5, 7, 8}

24. ∅ or { }

25. {0, 1, 2, 3, 5, 6, 7, 8}

26. {0, 1, 2, 6, 7, 8}

27. {0, 2, 4, 6}

28. {2, 6}

29. {0, 1, 3, 5, 7, 8}

30. {4}

31. {1, 7, 8}

32. 7

33. 5

34. As we are studying the basics of set theory, we use "inclusive or."

35. You find the elements that are in both sets.

36. You join the two sets together by finding all the elements that are in at least one of the sets.

2.4 Mickey Mouse Problems (Venn Diagrams)

Types of Venn Diagrams

Venn diagrams are representations of sets that use pictures. We will work with Venn diagrams involving two sets and three sets.

In order to ease a discussion of Venn diagrams, we can identify each distinct region within a Venn diagram with a label, as shown in the table below. Our labels are actually the Roman Numerals for the numbers one through four (in the two-set diagram) and the numbers one through eight (in the three-set diagram):

Two-Set Venn Diagram	Three-Set Venn Diagram
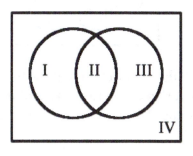	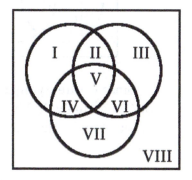

The concepts of **intersection, union,** and **complement** (as well as the corresponding notation) are also used in Venn diagrams, as we can shade parts of a diagram to represent a certain set. Furthermore, any time parentheses are involved in a notation statement, that piece of the corresponding Venn diagram is shaded first.

The rest of this section contains several examples, but the concepts of this section will be learned primarily through practice.

Two-Set Venn Diagrams

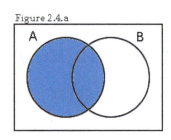

Figure 2.4.a

To represent the set A in a two-circle Venn diagram, simply shade the circle corresponding to set A, and ignore the rest of the figure, as shown in Figure 2.4.a.

Likewise, the same idea would be followed to indicate set B.

EXAMPLE 1: Create a Venn diagram for the set A ∩ B.

While we may be able to visualize the set A ∩ B in the diagram right away, let's take a look at a step-by-step approach. That process looks a little bit like making a cartoon strip.

To represent the set A ∩ B in a two-set Venn diagram, start by shading the set A. We will do this with vertical lines, as shown in Figure 2.4.b. Next, shade set B with horizontal lines, as shown in Figure 2.4.c. The crosshatched, football-shaped region in the center represents where the shadings overlap.

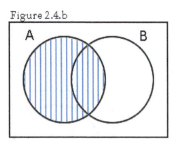
Figure 2.4.b

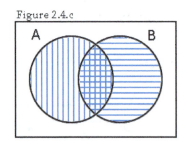
Figure 2.4.c

Since we are trying to shade ONLY the intersection of these two sets, we finish our diagram by darkly shading the overlap of the two sets, and erasing the parts of sets A and B that are not in the overlap. This is shown in Figure 2.4.d.

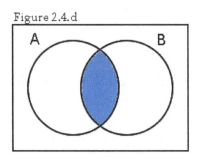
Figure 2.4.d

EXAMPLE 2: Create a Venn diagram for the set. A' ∪ B.

Again, we may be able to visualize the set A' ∪ B and draw the diagram right away. This is just fine, but for some the step-by-step approach will be helpful.

Start by lightly shading the set A' with vertical lines, as shown in Figure 2.4.e. Next, shade set B with horizontal lines. Like before, the crosshatched region represents where the shadings overlap, as shown in Figure 2.4.f.

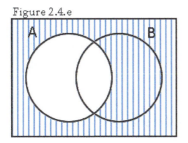

Figure 2.4.e

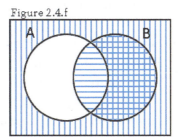

Figure 2.4.f

Since we are trying to shade the union of these two sets, we finish our diagram by joining the sets together. That is, we darkly shade everything we shaded in the previous image. This is shown in Figure 2.4.g.

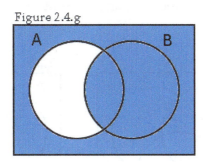

Figure 2.4.g

Three-Set Venn Diagrams

Just like we saw with two-set Venn diagrams, to represent the set A in a three-circle Venn diagram, we simply shade the circle corresponding to set A, and ignore the rest of the figure, as shown in figure 2.4.h. Likewise, the same idea would be followed to indicate set B or set C. (By the way, can you see why some people refer to these as Mickey Mouse problems?)

The creation of more involved three-set Venn diagrams is very similar to the process we followed for two-set Venn diagrams. They just take a little more time.

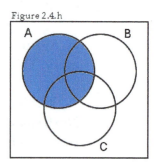
Figure 2.4.h

When considering the Venn diagram representation for the union and/or intersection of more than two sets, just like with the arithmetic order of operations, we need to work from left to right. And, once again, similar to the arithmetic order of operations, the only time we make an exception to the left-to-right process is when parentheses are used to group a specific operation.

Just like the arithmetic expressions "10 − 2 + 5" and "10 − (2 + 5)," simplify to two different values, the sets "A ∩ B ∪ C" and "A ∩ (B ∪ C)" will yield two different Venn diagrams.

EXAMPLE 3: Create a three-set Venn diagram for the set A ∩ B ∪ C.

Again, some of us may be able to visualize all or part of this diagram right away. If that is true, great. If not, fall back on a step-by-step approach.

Start by shading the set A with vertical lines, as shown in Figure 2.4.i. Next, shade set B with horizontal lines, as shown in Figure 2.4.j. Once again, the cross-hatched, football-shaped region represents where the shadings overlap.

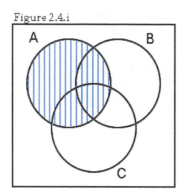
Figure 2.4.i

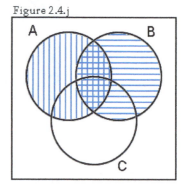
Figure 2.4.j

We're not done yet; we've only shown the first half. At this point we use that overlap and have shaded the set A ∩ B, as shown in Figure 2.4.k. Then we have to show the union of that football-shaped region with set C. So, we let the previously determined region be shaded with vertical lines, and then shade set C with horizontal lines, as shown in figure 2.4.l.

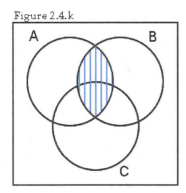

Figure 2.4.k

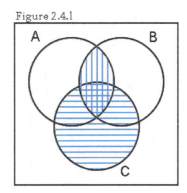

Figure 2.4.l

Finally, since we are trying to shade the union in this last step, we finish our diagram by joining the sets together. That is, we simply shade everything that was shaded in the previous image. The final image looks like Figure 2.4.m.

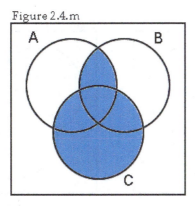

Figure 2.4.m

EXAMPLE 4: Create a three-circle Venn diagram for the set A ∩ (B ∪ C).

This time, because of the parentheses, we have to consider (B ∪ C) first. Thus, start by shading set B, as shown in Figure 2.4.n. Next, we shade set C, as shown in Figure 2.4.o. Once again, the crosshatched region represents where the shadings overlap. Since we want the union of these two sets, we now join the sets together and have shaded B ∪ C.

Figure 2.4.n

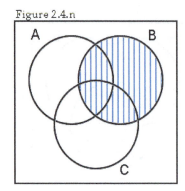

Figure 2.4.o

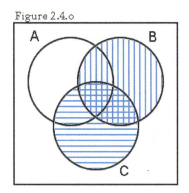

Next, we need to find the intersection of B ∪ C and set A. So, we take the horizontal shading of B ∪ C, as shown in Figure 2.4.p, and then shade set A with vertical lines, as shown in Figure 2.4.q.

Figure 2.4.p

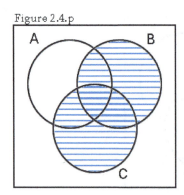

Figure 2.4.q

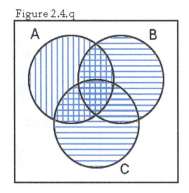

Finally, since we are trying to shade the intersection, in this last step, we finish our diagram by shading only the overlap from the previous image. The Venn diagram should look like Figure 2.4.r.

Figure 2.4.r

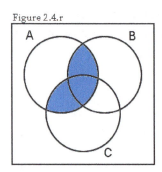

EXAMPLE 5: Create a three-set Venn diagram for the set B′ ∩ (A ∪ C).

Because of the parentheses, we must consider A ∪ C first. So, start by shading set A, as shown in Figure 2.4.s. Next, we shade set C, as shown in Figure 2.4.t.

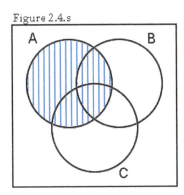

Figure 2.4.s

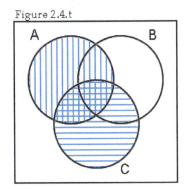

Figure 2.4.t

Since we want A ∪ C, we join the sets together, as shown in Figure 2.4.u. Next, we shade the set B′, as shown in Figure 2.4.v.

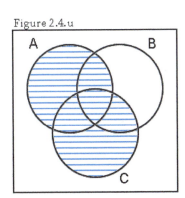

Figure 2.4.u

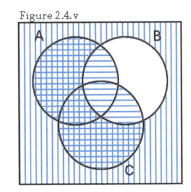

Figure 2.4.v

Finally, since we are trying to shade the intersection in this last step, we finish our diagram by shading only the overlap from the previous image. The final image looks like Figure 2.4.w.

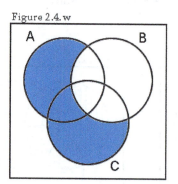

Figure 2.4.w

Section 2.4 Exercises

1. Given that

   P = {people who like prunes}

   Q = {people who like oranges}

   R = {people who like raspberries}

   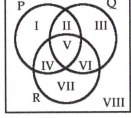

   Use the three-set Venn diagram shown to determine the region(s) corresponding to the following.

   a. Identify the region you belong in.

   b. People that like raspberries and oranges.

   c. People that like raspberries and oranges, but not prunes.

   d. People that like raspberries.

   e. People that like raspberries, but not oranges.

2. Create and shade a two-set Venn diagram for each of the following.

   a. A ∩ B

   b. A ∪ B

c. $A' \cap B$

d. $A \cup B'$

3. Create and shade a three-set Venn diagram for each of the following.

a. $A \cap C$

b. $B \cup C$

c. $B' \cap C$

d. $A \cup B'$

e. $A \cup (B \cap C)$

f. $B \cap (A \cup C)$

g. $(A \cup B) \cap C'$

h. $B \cup (A' \cap C)$

4. Use set notation to describe the Venn diagrams shown below.

a.
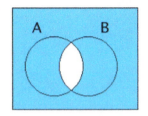

b.
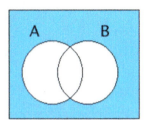

c.
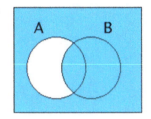

d.
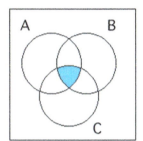

e.
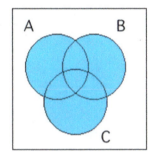

f.
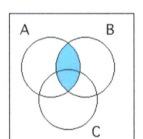

g.
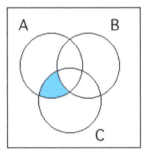

h.
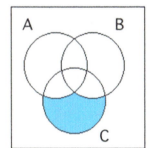

5. When drawing Venn diagrams, why do we use circles and not rectangles?

Answers to Section 2.4 Exercises

1. a. Answers will vary depending upon your personal preferences.

   b. Regions V and VI

   c. Region VI

   d. Regions IV, V, VI, and VII

   e. Regions IV and VII

2. a. 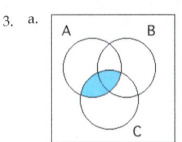   b.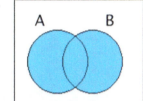

   c. 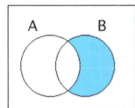   d.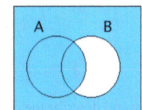

3. a. 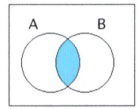   b.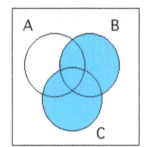

   c. 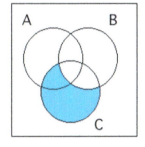   d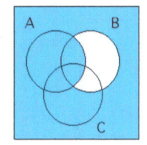

e.

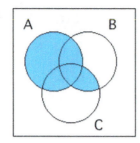

f.

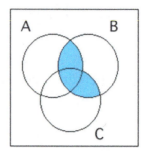

g.

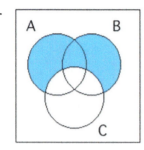

h.

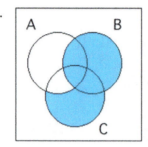

4. a. (A ∩ B)′ or A′ ∪ B′   b. (A ∪ B)′ or A′ ∩ B′
    c. A′ ∪ B   d. A ∩ B ∩ C

    e. A ∪ B ∩ C   f. A ∩ B

    g. A ∩ C ∪ B′   h. (A ∪ B)′ ∩ C

5. It is easier to identify and describe the different regions with circles. Try drawing a three-set diagram with rectangles, and you'll find it may quickly turn into a situation where it is difficult to tell where one rectangle ends and a different one begins.

2.5 Putting Mickey to Work (Applications of Venn Diagrams)

Applications of Venn Diagrams

In addition to just being fun, we can also use **Venn diagrams** to solve problems. When doing so, the key is to work from the "inside out," meaning we start by putting information in the regions of the diagram that represent the intersections of sets.

Pay attention to the fundamental difference between the last section and the coming material. The previous material was limited to identifying the various regions in a Venn diagram. Now, we will be counting how many items are in those regions.

EXAMPLE 1: In a group of 100 customers at Big Red's Pizza Emporium, 80 of them ordered mushrooms on their pizza, and 72 of them ordered pepperoni. 60 customers ordered both mushrooms and pepperoni on their pizza.

a. How many customers ordered mushrooms but no pepperoni?

b. How many customers ordered pepperoni but no mushrooms?

c. How many customers ordered neither of these two toppings?

Solution

Create a Venn diagram with two sets: mushrooms and pepperoni. To do this, first draw and label two intersecting circles inside a rectangle. Then, identify the number of items in each region, working from the inside out.

The innermost region is the football-shaped section corresponding to the intersection of the two sets. Since 60 customers ordered *both* mushrooms *and* pepperoni on their pizza, we begin by putting 60 in that center region.

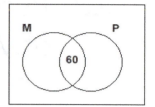

Next, we know we need to have a total of 80 customers inside the "M" circle. We already have 60 of them in there, so we have to put 20 more of them into the circle for set M, making sure they are also NOT in the circle for set P.

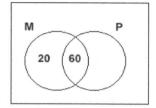

Similarly, we know we need a total of 72 customers inside the "P" circle. We already have 60 of them in there, so we need to put 12 more into the circle for set P, making sure they are *not* in the circle for set M.

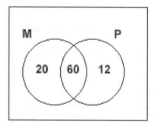

Finally, there are supposed to be 100 people represented in our diagram. Up to this point we have accounted for 92 of them $(20 + 60 + 12)$, so the remaining 8 customers must go into region outside both of the circles, but still inside our rectangle.

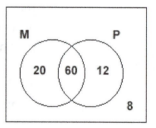

Now, we can answer the questions.

a. 20 ordered mushrooms but not pepperoni.

b. 12 ordered pepperoni but not mushrooms.

c. 8 ordered neither of these two toppings.

EXAMPLE 2: At Dan's Automotive Shop, 50 cars were inspected. 23 of the cars needed new brakes, 34 needed new exhaust systems, and 6 needed neither repair.

    a. How many cars needed both repairs?

    b. How many cars needed new brakes, but not a new exhaust system?

Solution

Create a Venn diagram with two sets. To do this, first draw two intersecting circles inside a rectangle. Be sure to label the circles accordingly.

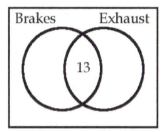

Now, work from the inside out. That is, begin by determining the number of cars in the intersection of the two sets.

Since 6 out of the 50 cars needed no repairs, leaving 44 cars that did need repairs. 23 needed brakes, and 34 needed exhaust systems. That makes 57 cars (23 + 34) that got worked on, which is too many; we know only 44 cars were worked on. This means 13 cars (57 − 44) got counted twice, which means that 13 cars get placed into the overlapping part of the Venn diagram (the intersection). These 13 cars needed both brakes AND exhaust systems.

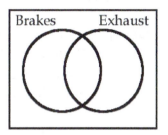

Look at the circle that corresponds to brakes. There should be 23 cars inside that circle. 13 are already accounted for, so the remaining 10 must be added into the brakes circle, but are outside of the exhaust circle. Likewise, 34 vehicles must appear in the exhaust circle, so 21 more must be placed inside that circle, but not that in the brakes circle.

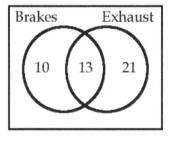

Finally, 6 cars need to be indicated outside the circles but inside the rectangle.

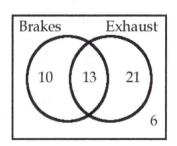

By looking at the completed Venn diagram, answer the original questions.

a. 13 cars needed both repairs.

b. 10 cars needed brakes, but not an exhaust system.

Applications with three-circle Venn diagrams are a bit longer and, consequently, a bit more involved. However, the strategy remains the same—work from the inside out.

EXAMPLE 3: A survey of 85 students asked them about the subjects they liked to study. 35 students liked math, 37 liked history, and 26 liked physics. 20 liked math and history, 14 liked math and physics, and 3 liked history and physics. 2 students liked all three subjects.

    a. How many of these students like math or physics?

    b. How many of these students didn't like any of the three subjects?

    c. How many of these students liked math and history but not physics?

Solution:

Create a Venn diagram with three sets, and label the circle M for math, H for history, and P for physics. Then, be sure to work from the inside out.

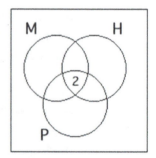

Start by placing the two students that like all three subjects into the center, which is the part of the diagram that represents the intersection of all three sets.

We know 20 students like math and history, so the intersection of those two sets must contain 20 students. We already have 2 of them in that intersection, so we put the remaining 18 in the intersection of the M and H circles, but not in the portion that also intersects the P circle.

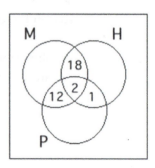

Using similar reasoning, we put 12 students and 1 student into the regions shown on the diagram.

Next, we know we need to have a total of 35 students inside the M circle. We already have 32 in there, so we put 3 students into Region I—the part of the M circle that does not intersect with any other region.

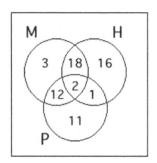

Similarly, we put 16 students into remaining section of the H circle and 11 students into the remaining section of the P circle.

Finally, there are supposed to be 85 students included in our diagram. Up to this point, we have included 63 of them, so the remaining 22 students must go into the portion of the diagram that is outside all of the circles, but still in the rectangle.

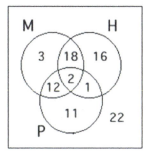

Now, we can answer the original questions:

a. 47 of the students like math or physics.

b. 22 of the students didn't like any of these subjects.

c. 18 of the students liked math and history but not physics.

Section 2.5 Exercises

1. The enrollment of 80 students at a college was examined. It was determined 39 students were taking a math class, 44 of them were taking an English class, and 18 of them were taking a math and an English class.

    a. How many students are not taking an English class?

    b. How many students are not taking a math class?

    c. How many students are taking neither of these types of classes?

    d. How many students are taking a math class or an English class?

2. A survey of 145 people was conducted, and the following data were gathered: 94 people use the Internet to find out about the news, 85 people use television to find out about news, and 62 people use the internet and television to find out about news.

    a. How many of the people use only television to find out about news?

    b. How many of the people do not use the Internet to find out about news?

    c. How many of the people do not use television to find out about news?

    d. How many of the people use the Internet or television to find out about news?

3. 120 people participated in a survey about their finances, and the following data were collected: 83 people have a checking account, 51 people have a savings account, and 26 have stocks. Forty people have a checking and a savings account, 11 have a checking account and stocks, and 7 have a savings account and stocks. Three people have all three types of accounts.

    a. How many people have only a savings account?

    b. How many people surveyed have none of the three types of accounts?

    c. How many people have a checking account and stocks, but do not have a savings account?

    d. How many people have only one of the three types of accounts?

4. A survey of 105 sports fans was conducted to determine what magazines they read. Fifty-two read *Sports Illustrated*, 43 read *ESPN The Magazine*, and 38 read *The Sporting News*. Eighteen read *Sports Illustrated* and *ESPN The Magazine*, 11 read *Sports Illustrated* and *The Sporting News*, and 8 read *ESPN The Magazine* and *The Sporting News*. Six people read all three magazines.

    a. How many of these sports fans read only *The Sporting News?*

    b. How many of these sports fans did not read any of these magazines?

    c. How many of these sports fans read *Sports Illustrated* or *The Sporting News?*

    d. How many of these sports fans read *ESPN The Magazine* and *The Sporting News* but do not read *Sports Illustrated?*

5. A survey of 160 people at a casino was conducted to determine the casino games that they typically play, and the following data were collected: Ninety-three played blackjack, 77 played roulette, and 39 played craps. Fifty-one played blackjack and roulette, 22 played blackjack and craps, and 11 played roulette and craps. Nine people played all three games.

    a. How many of these people played only craps?

    b. How many of these people did not play craps?

    c. How many of these people played at least one of the three games?

    d. How many of these people played blackjack or roulette?

6. A survey was conducted using 88 people. Of these people, 63 liked drinking coffee, 27 liked drinking tea, and 11 did not like either of these beverages. How many of the people surveyed liked both of these beverages?

7. The students in Mr. Jones' 5th grade class who owned pets filled out a survey. The Venn diagram below shows the number of students who belong to each region, with set D representing students who own dogs, set C representing students who own cats, and set F representing students who own fish.

    a. How many students own dogs?

    b. How many students own cats or fish?

    c. How many students own cats and fish?

    d. How many students do not own dogs, cats, or fish?

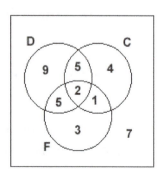

8. The provided Venn diagram represents the cardinality of the specific disjoint regions. Use that information to find the following cardinalities:

   a. n (A)

   b. n (C)

   c. n (A′)

   d. n (B ∩ C)

   e. n (A ∪ B′)

   f. n ((A ∪ C)′)

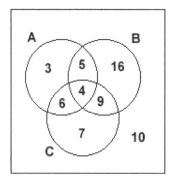

Answers to Section 2.5 Exercises

1. a. 36   b. 41   c. 15   d. 65

2. a. 23   b. 51   c. 60   d. 117

3. a. 7   b. 15   c. 8   d. 53

4. a. 25   b. 3   c. 79   d. 2

5. a. 15   b. 121   c. 134   d. 119

6. 13

7. a. 21   b. 20   c. 3   d. 7

8. a. 18   b. 26   c. 42   d. 13   e. 35   f. 26

Credits

1. Copyright © RupertMillard (CC by 3.0) at http://commons.wikimedia.org/wiki/File:Venn%27s_four_ellipse_construction.png.
2. Copyright © Wikimedia Foundation, Inc. (CC BY-SA 3.0) at http://en.wikipedia.org/wiki/John_Venn.
3. Copyright © New World Encyclopedia (CC BY-SA 3.0) at http://www.newworldencyclopedia.org/entry/John_Venn.
4. "Venn John signature," http://commons.wikimedia.org/wiki/File:Venn_John_signature.jpg. Copyright in the Public Domain.
5. Copyright © (CC by 2.5) at https://upload.wikimedia.org/wikipedia/commons/4/49/Venn-stainedglass-gonville-caius.jpg.
6. Copyright © Wikimedia Foundation, Inc. (CC BY-SA 3.0) at http://en.wikipedia.org/wiki/Georg_Cantor.
7. "Georg Cantor," http://commons.wikimedia.org/wiki/File:Georg_Cantor2.jpg. Copyright in the Public Domain.
8. "Gold Ingots on white background," http://commons.wikimedia.org/wiki/File:Gold_Ingots_on_white_background.jpg. Copyright in the Public Domain.
9. "Pizza Slice," http://pixabay.com/en/pizza-food-slice-cheese-mushroom-23477/. Copyright in the Public Domain.
10. "Red Cardinal Bird," http://pixabay.com/en/cardinal-bird-animal-red-154622/. Copyright in the Public Domain.
11. "Oranges," http://pixabay.com/en/orange-oranges-fruit-sweet-food-207819/. Copyright in the Public Domain.
12. "Champignion," http://pixabay.com/en/champignion-healthy-mushroom-343958/. Copyright in the Public Domain.
13. "Mechanic," http://pixabay.com/en/mechanic-car-service-repair-346257/. Copyright in the Public Domain.

STATISTICS

I n statistics, we study **random systems**. We take samples, organize data, and draw conclusions. If, however, the systems we studied were not random, they would be completely predictable, and there would be no need for statistics, at all.

More and more businesses and other entities are employing statisticians. Through random, unbiased sampling, a statistician can save a company millions and better prepare them for upcoming events. Professional sports teams analyze players' performances in different situations, so the teams can increase their chances of winning. And, if one team is doing it, the others had better follow suit or risk falling behind even further.

Insurance companies employ actuaries to analyze data and help determine how much policyholders pay for insurance policies. A wealth of information is used to determine the likelihood a non-smoking, single mother of three has of getting a speeding ticket, having an accident, or even passing away. Based on that data, the insurance company sets its rates.

According to its website, http://www.nielsen.com/us/en.html, **The Nielson Company** provides timely data on media and consumer trends, including TV ratings, smartphone trends, and video game purchase intent. That data provides other companies with a better understanding of consumers, but it also paints a rich portrait of the American audience.

Statistics are all around us. If we have a better understanding of how the data were obtained and who interpreted the results, we can get a better idea of what to believe and, unfortunately, what not to believe.

Do be cautious, though. An old adage is "90% of all statistics can be made to say anything … 50% of the time."

3.1 On the Shoulders of Giants (Biographies & Historical References)

Quick! Think of a famous nurse. Many people respond to this prompt by naming Florence Nightingale. What you may not have realized is she became that famous nurse through the effective use of statistics.

On the other side of the statistics spectrum is an infamous photo. In 1948, Harry Truman defeated Thomas Dewey for the Presidency of the U.S. In a well-published photo, the day after the election, the President-Elect laughingly waved a Chicago Tribune with the headline that mistakenly read "DEWEY DEFEATS TRUMAN." George Gallup was the one ultimately responsible for that headline.

Florence Nightingale

Florence Nightingale (1820–1910) was a celebrated English social reformer and statistician, and the founder of modern nursing. She was born on May 12, 1820 into a rich, upper class, well-connected British family in Florence, Italy, and was named after the city of her birth. In her youth, she was respectful of her family's opposition to her working as a nurse, only announcing her decision to enter the field in 1844. Despite the intense anger and distress of her mother and sister, she rebelled against the expected role for a woman of her status, which was to become a wife and mother. In spite of opposition from her family, and the restrictive social code for affluent young English women, Nightingale worked diligently to educate herself in the art and science of nursing.

Florence Nightingale's most famous contributions came during the Crimean War, which became her central focus when reports about the horrific conditions for the wounded were sent back to Britain She, and the staff of 38 women volunteer nurses that she trained, were sent to care for the soldiers that were fighting in the Ottoman Empire. Her team found that an overworked medical staff, in the face of official indifference, was delivering poor care for wounded soldiers. Medicines were in short supply, hygiene was being neglected, and mass infections were common, many of them fatal.

After Nightingale sent a plea to *The Times* for a government solution to the poor condition of the facilities, the British Government commissioned the building of a prefabricated hospital that could be built in England and shipped to Southern Europe. When she first arrived, ten times more soldiers died from illnesses such as typhus, typhoid, cholera and dysentery than from battle wounds. Nightingale's directives in the war hospital, such as hand washing and other hygiene practices, helped reduce the death rate from 42% to 2%.

Nightingale became a pioneer in the visual presentation of information and statistical graphics. She is credited with developing a form of the pie chart, now known as the polar area diagram, to illustrate seasonal sources of patient mortality in the military field hospital she managed. She called a compilation of such diagrams a "coxcomb," and frequently used them to present reports on the nature and magnitude of the conditions of medical care. These visual representations of the data helped the Members of Parliament and civil servants to read and understand traditional statistical reports. In 1859, Nightingale was elected the first female member of the Royal Statistical Society. She later became an honorary member of the American Statistical Association.[1]

George Gallup

George Horace Gallup Jr. (1901–1984) was born in Jefferson, Iowa. As a teen, George would deliver milk, and he used his salary to start a newspaper at the local high school. Later he attended the University of Iowa, where he became the editor of The Daily Iowan, an independent newspaper that serves the university campus. He earned his B.A. in 1923, his M.A. in 1925 and his Ph.D. in 1928. After earning his doctorate, he moved to Des Moines, Iowa, where he served as head of the Department of Journalism at Drake University until 1931. Later that year, he moved to Evanston, Illinois, as a professor of journalism and advertising at Northwestern University. A year later, he moved to New York City to join the advertising agency of Young and Rubicam as director of research. He was also a professor of journalism

---

at Columbia University, but he had to give up this position shortly after he formed his own polling company, the American Institute of Public Opinion (Gallup Poll), in 1935.

Gallup is often credited as the developer of public polling. He wished to objectively determine the opinions held by the people. To ensure his independence and objectivity, Gallup resolved that he would undertake no polling that was paid for or sponsored, in any way, by special interest groups such as the Republican and Democratic parties.

In 1936, his new organization achieved national recognition by correctly predicting, from the replies of only 50,000 respondents, that Franklin Roosevelt would defeat Alf Landon in the U.S. Presidential election. This was in direct contradiction to a poll in the widely respected *Literary Digest* magazine. Twelve years later, his organization had its moment of greatest ignominy, when it predicted that Thomas Dewey would defeat Harry S. Truman in the 1948 election, by five to fifteen percentage points. Gallup believed the error was mostly due to ending his polling three weeks before Election Day.

Today, the Gallup Organization conducts 1,000 interviews per day, 350 days out of the year, using both landline and cell phones across the U.S. for its surveys. Their tracking methodology relies on live interviewers, dual-frame random-digit-dial sampling and uses a multi-call design to reach respondents not contacted on the initial attempt. The data are weighted daily by the number of adults in the household and the respondents' reliance on cell phones, to adjust for any disproportion in selection probabilities. The data are then further weighted to compensate for nonrandom nonresponse, using targets from the U.S. Census Bureau for age, region, gender, education, Hispanic ethnicity, and race. The resulting sample represents an estimated 95% of all U.S. households.[2]

---

2 Copyright © Wikimedia Foundation, Inc. (CC BY-SA 3.0) at http://en.wikipedia.org/wiki/George_Gallup.

3.2 Right Down the Middle (Mean, Median & Mode)

Measures of Central Tendency

The thing everyone loves to call the "average" is actually called the mean. Average is a general term that actually applies to one of several different measures of central tendency. A couple of these measures are the mean and the median.

The **mean** is commonly known as the "average," and this is what we get when we "add them all up and divide by how many there are." You are probably very familiar with the mean.

The **median** is the middle number in the data set. Be careful; the data values must be arranged in order before you can find the median. This order can be from lowest to highest, or from highest to lowest, but either way, the median is the value in the middle. For example, if our data set is {3, 3, 8, 7, 5}, we might be confused into thinking that the median is 8, because it appears in the middle. But remember, the numbers must be ordered first. So, since our ordered data set is actually {3, 3, 5, 7, 8} when arranged from lowest to highest, we state the median is 5.

What happens to the median if there is an even number of data values? The median is then found by finding the mean of the two numbers in the middle. For example, if the data set is {2, 3, 5, 7, 8, 9}, there is not a single middle number. Since 5 and 7 are the two numbers in the middle, the median would be the mean of those two numbers, or $5 + 7 = 12$, and $12/2 = 6$. Thus, state the median for that set would be 6.

Another descriptive measure of a data set is the mode. The **mode** is the single data value that occurs most often within the data set. If there are two data values that occur more often than the others then we refer to the set as **bimodal**. For example, the set of data is {2, 3, 4, 5, 7, 2, 7, 8, 2, 7, 7, 2} is bimodal, as the 2 and the 7 occur four times each. If there is not a single value (or a pair of values) that occurs most often, we say there is **no mode**. It is very common to see the mode grouped with the mean and median when discussing measures of central tendency, but since modal values don't have to appear around the middle of a data set, it is a misnomer to say they are describing the central tendencies of a set. That said, they are still useful in describing the overall appearance of the set.

EXAMPLE 1: Find the mean, median, and mode of the exam scores listed below. Round your answers to the nearest tenth, as necessary.

93, 97, 59, 71, 57, 84, 89, 79, 79, 88, 68, 91, 76, 87, 94, 73, 58, 85, 82, 38

For the mean, the sum of the numbers is 1,548, and 1,548/20 = 77.4

For the median, first order the numbers from lowest to highest. Since there is an even number of values, we must find the mean of the two in the middle. (79+82)/2 = 80.5.

The mode is 79, as it is the most common value.

A **frequency table** is a way to arrange all of the data values from a given situation into chart form. Here is an example of a frequency table showing the size of the litter for 23 different cats.

Number of Kittens in Litter	Frequency
1	2
2	5
3	6
4	7
5	3

The chart above represents the following data set: {1, 1, 2, 2, 2, 2, 2, 3, 3, 3, 3, 3, 3, 4, 4, 4, 4, 4, 4, 4, 5, 5, 5}, but using the chart allows us to avoid writing it out. We could certainly list out all 23 data values to find the mean, median, and mode, but let's take a different and more efficient approach.

The Mean: To find the mean, we need to add the data values together and, then, divide by the total number of data values. Adding them up, we have: $(1 \times 2) + (2 \times 5) + (3 \times 6) + (4 \times 7) + (5 \times 3) = 73$. Since there are 23 total data values, the mean (rounded to the tenth) will be $73/23 = 3.2$.

The Median: Since we have 23 data values, an odd number, there will be a "middle number." There will be eleven data values above the median and eleven data values below the median. Using the chart to our advantage, we can just count up from the bottom and find the location of the 12th data value. Counting the three 5's and the seven 4's gets us to the 10th data value. Since there are six 3's, we can see that the 12th data value would be a 3, and thus, 3 is the median. If we preferred, we could have counted down from the top to find the 12th data value. Counting the two 1's and the five 2's gets us to the 7th data value. Since there are six 3's, the 12th data value will be in that group, and the median will (again) be 3.

The Mode: The mode is the value that occurs most often and is easily found by looking at the table. There are seven 4's, which is the most common of the data values. So, the mode is 4.

EXAMPLE 2: Create a frequency table for the data set {2, 3, 4, 5, 7, 2, 7, 8, 2, 7, 7, 2, 8, 5, 5, 5, 5}, and then find the mean, median and mode, rounding to the tenth, as necessary.

First, create the frequency table. Be sure to order the data.

Data Value	Frequency
2	4
3	1
4	1
5	5
7	4
8	2
Total:	17

Mean: Multiply each data value by its frequency. Add up those products, and divide by the total number of data values. $(2 \times 4 + 3 \times 1 + 4 \times 1 + 5 \times 5 + 7 \times 4 + 8 \times 2)/17 \approx 4.9412$. Thus, to the tenth, the mean is 4.9

Median: Since there are 17 data values, the median will be the one in the ninth position. There are four 2s, one 3, and one 4. That makes six values less than 5, so the ninth data value will be the third 5. Thus, the median is 5.

Mode: The value of 5 occurs most often, so the mode is 5.

Grade Point Averages

Students need to be able to compute their own **Grade Point Average,** also known as the **GPA**. The GPA is a weighted average. That is, each grade has a weight associated with it. This weight is typically the number of credits of the course for which the grade was earned. Thus, an A in a 4-credit class carries more weight than an A in a 3-credit class.

First, we need to know the values of specific grades. The following table is taken from the College of Southern Nevada (CSN) Catalog.

Grade	Value	Grade	Value
A	4.0	C	2.0
A–	3.7	C–	1.7
B+	3.3	D+	1.3
B	3.0	D	1.0
B–	2.7	D–	0.7
C+	2.3	F	0.0

Courses with grades of I (Incomplete), P (Pass), S (Satisfactory), U (Unsatisfactory), W (Withdrawal), NR (Not Reported), or AU (Audit) are not included in the GPA calculation.

To compute a GPA, we first multiply the value of the grade by the corresponding weight, which is the number of credits for the course. That product gives us the number of grade points for the course. Then, we add together the Grade Points for all the courses and divide by the total number of credits taken. At CSN (and most schools), published GPA's are usually rounded to the nearest hundredth of a point, eventhough individual grade values are listed to the tenth of a point.

EXAMPLE 3: The table below represents a student's grades for a given semester at CSN. Find the student's GPA for that semester. Notice that the student took a total of 12 credits.

Course	Credits	Grade
ENG 103	3	A–
HIST 107	4	B
MATH 120	3	C+
IS 241	2	A

For ENG 103, $3 \times 3.7 = 11.1$
For HIST 107, $4 \times 3.0 = 12.0$
For MATH 120, $3 \times 2.3 = 6.9$
For IS 241, $2 \times 4.0 = 8.0$

Semester GPA: $(11.1 + 12.0 + 6.9 + 8.0)/12 = 38.0/12 = 3.16666 \ldots$, which is then rounded to 3.17.

Section 3.2 Exercises

For Exercises #1 through #4, find the mean, median, and mode of each data set. Round any decimal answers to the nearest tenth.

1. {1, 3, 5, 5, 7, 8, 9, 11}

2. {5, 13, 8, 4, 7, 2, 11, 15, 3}

3. {37, 52, 84, 99, 73, 17}

4. {4, 16, 9, 4, 2, 4, 1}

5. During the month of November 2008, the Oakland Raiders played five football games. In those games, the team scored the following number of points: 0, 6, 15, 31, and 13. Find the mean, median, and mode for this data set. Round any decimal answers to the nearest tenth. Did the Raiders win all five games?

6. The mean score on a set of 15 exams is 74. What is the sum of the 15 exam scores?

7. A class of 12 students has taken an exam, and the mean of their scores is 71. One student takes the exam late and scores 92. After including the new score, what is the mean score for all 13 exams? If you get a decimal answer, you should round to the nearest hundredth.

8. The 20 students in Mr. Edmondson's class earned a mean score of 76 on an exam. Taking the same exam, the 10 students in Mrs. Wilkinson's class earned a mean score of 86. What is the mean when these teachers combine the scores of their students? If you get a decimal answer, you should round to the nearest hundredth.

9. After six exams, Carl has a mean score of 78.5. With only one exam remaining in the class, what is the minimum score Carl will need on that exam to have an overall mean of 80?

10. Create a set of seven data values in which the mean is higher than the median.

11. Can the mean be a negative number? Explain your answer, and give an example.

12. The table at the right gives the ages of cars (in years) in a supermarket parking lot. Using the information given in the table, find the mean, median, and mode of the data. Round any decimal answers to the nearest tenth.

Age of the Car in Years	Number of Cars
1	5
2	9
3	13
4	14
5	6
6	2

13. The following table gives the distance (in miles) students in a class travel to get to campus. Using the information given in the table, find the mean, median, and mode of the data. Round any decimal answers to the nearest tenth of a mile.

Miles Traveled	Number of Students
2	4
5	8
8	5
12	6
15	3
20	2
30	1

14. Given the list of courses and grades, calculate the GPA of this student. For grade values, use the CSN values listed earlier in this section. Round your answer to the nearest hundredth.

Course	Credits	Grade
Chemistry	5	A
Art	1	C
Spanish	3	B

15. Given the list of courses and grades, calculate the GPA of this student. For grade values, use the CSN values listed earlier in this section. Round your answer to the nearest hundredth.

Course	Credits	Grade
Math	3	C+
English	3	B
History	4	C
PE	1	A

16. Would it make sense to have a frequency table for a set of data without repeated values? Why or why not?0

Answers to Section 3.2 Exercises

1. mean = 6.1, median = 6, mode = 5

2. mean = 7.6, median = 7, mode = none

3. mean = 60.3, median = 62.5, mode = none

4. mean = 5.7, median = 4, mode = 4

5. mean = 13, median = 13, mode = none
   Since they could not win a game in which they scored zero points, they did not win all five games.

6. 1,110

7. 72.62

8. 79.33

9. 89

10. Answers may vary. Find the mean and median to check and see if your data set meets the conditions of the problem.

11. Yes. An example would be the mean low temperature in Anchorage, Alaska, during the month of January.

12. mean = 3.3, median = 3, mode = 4

13. mean = 9.5, median = 8, mode = 5

14. 3.44

15. 2.54

16. No. Without repeated values, you would end up listing all the data values. In that case, just leave them in a set, rather than building a table to indicate there is only one of each value.

3.3 Mine is Better Than Yours (Percentiles & Quartiles)

Percentiles

When data sets are relatively small, one of the best ways to spot clusters of values, as well as the relationship between the individual values, is to simply list and inspect them. However, when the data sets get large, say over 100 values, listing the values out becomes impractical. As an alternative, **percentiles** are used to describe the relative position of certain data values in relation to the whole data set. Simply put, a data value at the 70th percentile is above 70% of the data values. We hear the term **percentile rank** often when discussing test results. If a student scores at the 82nd percentile, then he scored above 82% of the people that took the test.

Read that last sentence again. If a student scores at the 82nd percentile, then he scored *above* 82% of the people that took the test. The biggest mistake when dealing with percentiles is computing them as straightforward percents. They are actually a little bit different. If we have a set of 10 ranked values, 70% of them would be 7 out of the 10. However, the value at the 70th percentile is the first value greater than 70% of the values in the set. So, if the data consists of the set {1, 2, 3, 4, 5, 6, 7, 8, 9, 10}, the score at the 70th percentile is the 8, as it is the first value in the set *higher than* 70% of the vales in the set. Once again, in most cases, if a mistake is made in working with percentiles, it is finding the value *at* the stated percent, rather than finding the first value *above* the stated percent.

Next, we need to make sure whether we are looking for an overall rank or a percentile rank. Overall rank is based on the total number of data values, and the highest value is ranked first. Percentile ranks are from 0 through 99. There is no 100th percentile, as a value at the 100th percentile would be above 100% of the values in the set, including the value itself. In other words, since a value in the set cannot be higher than itself, it cannot be higher than all the points in the set.

EXAMPLE 1: In a class of 375 students, Sarah has a rank of 15th. What is her percentile rank?

We first find the number of students below her: $375 - 15 = 360$. We then divide that number by the total number of students in the class: $360/375 = 0.96$. So, Sarah is ranked at the 96th percentile. This means she is ranked above 96% of the students in her class.

EXAMPLE 2: 415 people ran a marathon, and Shane finished at the 27th percentile. In what place did Shane finish?

We use the knowledge that he finished ahead of 27% of the runners to determine that he finished ahead of $0.27 \times 415 = 112.05 = 112$ runners. Since 415 runners were in the race, we can subtract the 112 that finished behind him: $415 - 112 = 303$, to find out that he finished in 303rd place.

Quartiles

Instead of discussing the position of a data value in relation to all the values in the set, an alternative descriptive measure used in large data sets is the inspection of quartiles. **Quartiles** cut the data into four equal parts. As we saw in the previous section, the middle of the ranked data set is called the **median.** The median cuts the data set in half. The median does not have to be a data value. In fact, if there is an even number of data values, the median is the mean of the two values in the middle. Thus, the median can be thought of as a boundary between the lower half and the upper half of the data set.

Counting from the bottom up, the **first quartile** (also called the **lower quartile**) is the median of the lower half of the data set. Similarly, the **third quartile** (also called the **upper quartile**) is the median of the upper half of the data set. Like the median (sometimes called the **second** or **middle quartile**) of the entire set, the first and third quartiles do not have to be specific values in the data set. The first quartile, median, and third quartile are boundaries that divide the data set into four equal parts. Remember, percentiles describe the location of a specific data values, and quartiles cut the entire set into four equally sized parts.

Once again, although it is possible to have data values fall at the quartile boundaries, it is not necessary. It is also worth noting that, when considering an odd number of data points, the median is *not* in either the upper or lower half; it is the boundary between the two. Thus, in the set {1, 2, 3, 4, 5, 6, 7}, the lower half of data points is {1, 2, 3}, not {1, 2, 3, 4}.

The Five-Number Summary

When working with quartiles, we should also identify the lowest and highest values in the data set. All together, that group of five quantities—the lowest value (L), the first quartile (Q_1), the median (M), the third quartile (Q_3), and the highest value (H)—is called the **five-number summary**.

EXAMPLE 3: Indicate the five-number summary for the data set {35, 50, 51, 55, 56, 58, 62, 62, 66, 72, 74, 74, 74, 77, 81, 87, 90, 95, 95, 99}

The easy ones are the lowest and highest values: L = 35 and H = 99.

To find the first and third quartiles, we have to find the median first. Of the 20 values, the 72 and the first 74 are in the middle. Thus, M = (72 + 74)/2 = 73.

Then, for the first quartile, Q_1, consider only the lower half of the data set. Of those 10 values, the 56 and 58 are in the middle. Thus, Q_1 = (56 + 58)/2 = 57.

Likewise, for the third quartile, Q_3, we consider only the upper half of the data set. Q_3 = (81 + 87)/2 = 84.

Putting it all together: L = 35, Q_1 = 57, M = 73, Q_3 = 84, and H = 99.

Box-and-Whisker Plots

The visual representation of the five-number summary is called a **box-and-whisker plot** (or **boxplot**). The five-number summary indicates the boundaries that cut the data set into four equal parts, each of which contains approximately 25% of the data points. The number of data values between the lowest value (L) and the first quartile (Q_1) is the same as the number of data values between the median (M) and the third quartile (Q_3).

Be careful with this. The length of the intervals, on the number line, may be (and is often) quite different from quartile to quartile, but remember, each interval contains approximately 25% of the values that are in the entire data set. Also, remember, we are usually dealing with large sets of data.

On a boxplot, the values appear in order from least to greatest with a "box" drawn in the center, using the first and third quartile values as the left and right sides of the box. The median is located somewhere inside the box, and the highest and lowest values are used to draw "whiskers" on the box.

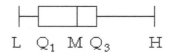

$$L \quad Q_1 \quad M \quad Q_3 \qquad H$$

The values of the five-number summary should always appear immediately above or below the boxplot. For indexing purposes, we will also occasionally see a number line drawn underneath the entire boxplot. Without the values of the five-number summary, the boxplot is just a strange-looking figure.

EXAMPLE 4: The following boxplot represents the amount of snowfall, in inches, recorded at 240 weather stations around a large city in a given week last December.

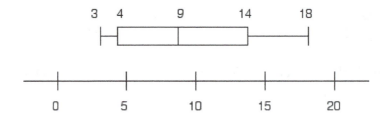

a. What was the median snowfall level?

b. Approximately how many weather stations recorded between 9 and 14 inches of snow?

c. Approximately how many weather stations recorded between 3 and 14 inches of snow?

d. What was the largest snowfall reading?

e. Did more weather stations record 3 to 4 inches or 4 to 9 inches of snow?

Answers

a. The median is the value indicated inside the "box" part. Thus, the median snowfall was 9 inches.

b. Since each vertical mark is a quartile label, approximately 25% of the values fell within each interval. Since there are 240 weather stations, 60 of them recorded 9 to 14 inches of snow.

c. From 3 to 14 inches, we cover three intervals, which is 75%. So, 180 weather stations recorded 3 to 14 inches of snow.

d. The largest snowfall reading is at the end of the rightmost whisker, which is 18 inches.

e. Remember, each interval represents 25% of the data values. Thus, although these two intervals do not appear to be the same length on the number line, they contain the same number of data points. The smaller sections (like between 3 and 4) indicate there are several data points squeezed into a small space, and larger sections (like between 4 and 9) show where the data points are more spread out.

EXAMPLE 5: The following boxplot represents the test results for the 560 students who took the CHEM 120 final exam at State University last fall.

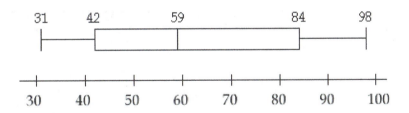

a. A student with a score of 84 did better than how many students?

b. What was the highest score?

c. Did more students score above 59 or below 59?

d. What percent of students scored lower than 42?

Answers

    a. 84 is at the third quartile, which is the 75th percentile. Thus, a student with a score of 84 did better than 420 (0.75 × 560) students.

    b. The highest score was 98.

    c. 59 is the median. Half the students scored above 59, and half scored below it.

    d. 42 is at the first quartile. So, 25% of the scores were below it.

Another practical use of boxplots is in the comparison of multiple sets. By stacking two or more boxplots onto the same graph, we can quickly compare the various aspects of the data at hand.

EXAMPLE 6: Walmart, K-Mart, and Target all sell a number of different CDs. The stack of boxplots below represents the costs of all the CDs at each of the stores.

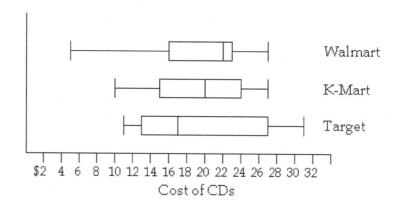

    a. How much is the least expensive CD at K-Mart?

    b. Which of the three stores sells the most expensive CD? How much is that CD?

    c. Which store has the lowest-priced CD? How much is that CD?

    d. Which store has the largest price range?

    e. Which store has the highest median price?

    f. For Target, are there more CDs priced above or below $17?

    g. For Walmart, are there more CDs priced above or below $20?

Answers

a. $10

b. Target has the most expensive CD, and it costs $31.

c. Walmart has the lowest priced CD, and it CD costs $5.

d. Walmart's price range of $22 is the largest.

e. Walmart has the highest median price.

f. Neither. $17 is the median, so half the CDs are priced above $17, and half are priced below $17.

g. Be careful. Do not look at the length of the boxplot above and below $20. Remember, a longer region on the boxplot just means the values are more spread out. Then, realize the median is $22. That means half the CDs are priced above $22. Thus, *more than half* of the CDs must be priced above $20.

Section 3.3 Exercises

1. In a graduating class of 300 students, Erin has a rank of 18th. What is her percentile rank in the class?

2. In a 13-mile half-marathon, Todd finished in 33rd place. There were 175 runners, what was Todd's percentile rank?

3. In a class with 35 students, Gil scored at the 60th percentile on an exam. How many students scored lower than Gil?

4. Crystal has a percentile rank of 20 in her graduating class, which consists of 400 students. What is her rank in the class?

5. Amy and Joe are in the same senior class. Amy is ranked 40th out of the 280 seniors, and Joe has a percentile rank of 15. Of these two seniors, which one is ranked higher?

6. In a class of 30 students, Bruce is ranked 29th. What is his percentile rank? In that same class, Jennifer is ranked 3rd. What is her percentile rank?

7. Karen is ranked at the 75th percentile in her class of 440 students. What is her rank in the class?

8. For any data set, approximately what percent of the data points are above the median?

9. For any data set, approximately what percent of the data points are above the first quartile?

10. For any data set, approximately what percent of the data points are between the first and third quartiles?

11. For the following boxplot, find the five-number summary.

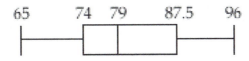

12. Use the following set of data to find the five-number summary.

12, 12, 13, 13, 13, 16, 16, 16, 16, 18, 18, 18, 21, 21, 21, 26, 26, 28, 31, 51, 51, 75

13. Use the following set of data to find the five-number summary.

   65, 65, 67, 68, 74, 74, 75, 77, 77, 77, 79, 80, 83, 85, 85, 86, 89, 92, 93, 94, 96

For Exercises #14 through #17, use the following boxplot, which represents number of points scored by 160 different basketball players during a tournament.

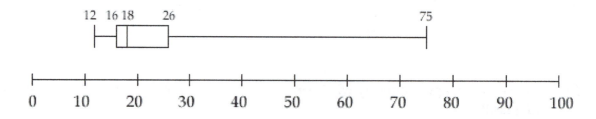

14. What was the highest score?

15. What was the median score?

16. Approximately how many players scored more than 26 points?

17. Were there more players who scored between 16 and 18 points or more who scored between 26 and 75 points?

For Exercises #18 through #20, use the following pair of boxplots, which indicate the cost per credit for community colleges in Indiana and Texas. Note, these numbers are for this exercise only, and may not be reflective of the actual costs.

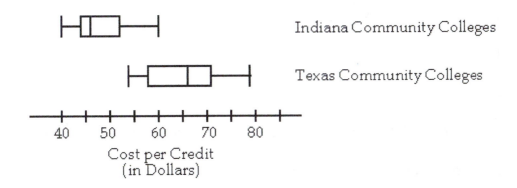

18. Estimate the values for the five-number summary for each state.

19. As a general statement, is it less expensive to attend a community college in Indiana or Texas?

20. Do the costs vary more in Indiana or Texas? How can you tell?

Answers to Section 3.3 Exercises

1. 94

2. 81

3. 21

4. 320th

5. Amy is ranked higher. Amy's percentile rank is 86. Joe's rank in the class is 238th.

6. Bruce's percentile rank is 3. Jennifer's percentile rank is 90.

7. 110th

8. 50%

9. 75%

10. 50%

11. $L = 65$, $Q_1 = 74$, $M = 79$, $Q_3 = 87.5$, $H = 96$

12. $L = 12$, $Q_1 = 16$, $M = 18$, $Q_3 = 26$, $H = 75$

13. $L = 65$, $Q_1 = 74$, $M = 79$, $Q_3 = 87.5$, $H = 96$

14. 75

15. 18

16. 40

17. There were 40 players in both those intervals.

18. IN: $L = \$40$, $Q_1 = \$44$, $M = \$45$, $Q_3 = \$52$, $H = \$60$;
    TX: $L = \$54$, $Q_1 = \$57$, $M = \$66$, $Q_3 = \$70$, $H = \$79$

19. Although it is possible to find some community colleges in Texas that are less expensive than some community colleges in Indiana, in general, the cost is less expensive in Indiana.

20. Costs vary more in Texas. We know this because the width of the box-and-whisker plot is larger for Texas.

3.4 Pretty Pictures (Graphs of Data)

Organizing Data

The following are the scores on an exam.

93, 97, 59, 71, 57, 84, 89, 79, 79, 88, 68, 91, 76, 87, 94, 73, 57, 85, 82, 38

A **stem-and-leaf plot** uses the first digit of the scores as the "stem" and the last digit of the scores as the "leaf." This method of organizing data helps us put the numbers in order, from lowest to highest. For the exam scores listed above, we create the necessary "stems" first. Our data set ranges from the 30s to the 90s, so we need to establish the stems between 3 and 9 for that range of scores.

Stem	Leaves
9	
8	
7	
6	
5	
4	
3	

We now put on each of the data points, using the ones digits as the "leaves." Keep in mind, the 3 in the top row of the Leaves column represents the number 93, because it uses the 9 in the stem column as the tens digit. In similar fashion, the 8 in the bottom row of the Leaves column represents the number 38.

Stem	Leaves
9	3 7 1 4
8	4 9 8 7 5 2
7	1 9 9 6 3
6	8
5	9 7 7
4	
3	8

Finally, we will finish our stem-and-leaf plot by rearranging the numbers in each of the leaves from lowest to highest. Now we have a very nice chart that lists all of the data points in order from lowest to highest.

Stem	Leaves
9	1 3 4 7
8	2 4 5 7 8 9
7	1 3 6 9 9
6	8
5	7 7 9
4	
3	8

Recall, a frequency table is a way to arrange all of the data from a given situation into chart form. Using the exam scores from before, here is an example of a frequency table, where the data has been grouped into categories that we have deemed to be important. The Frequency column lists us the number of data points included within each specific category.

Grade	Frequency
A	4
B	6
C	5
D	1
F	4
Total:	20

Notice the last row in the table. It represents the total number of data values. The total number of data values is important when we examine the percent of the data in each category. That will lead us to a relative frequency table. But first, it is important to examine the individual frequencies of the specific grades.

A **bar graph** (also called a **bar chart**) is a visual way to present categories of data. Each of the categories (also called **bins**) is typically listed across a horizontal axis, and the number of items in each category (i.e., the frequency) is indicated by the height of the corresponding bars. In a bar graph, we also see spaces between the bars to indicate each category is separate and distinct. The bars on a bar graph can be horizontal, but we will stick with a vertical representation.

EXAMPLE 1: A bag contains four red jellybeans (R), six green jellybeans (G), two purple jellybeans (P), seven yellow jellybeans (Y), and four blue jellybeans (B). Create a bar graph that displays this data.

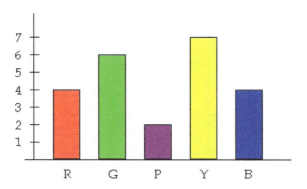

A **histogram** is similar to a bar graph. The big difference is that the categories displayed on the horizontal axis represent a *continuous* set of data. These categories are usually based on percents or time, and to show there are no gaps in the data, there are no spaces in between the bars.

EXAMPLE 2: The following table represents the number of babies born in a hospital's maternity ward. Create a histogram that represents the data.

Year	Babies
2001	981
2002	1895
2003	2027
2004	651
2005	432

144

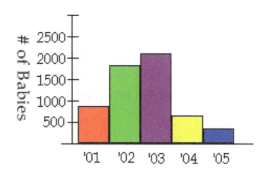

Appearance-wise, histograms and bar charts look very similar. Remember, bar charts show data sets that are independent of each other and can make sense in any order. Histograms show ordered data and are designed to have no spaces between the bars to reflect their data of continuous measurement. For example, in the histogram for the babies shown above, it would not make sense to put the categories—which represent consecutive years—in a different order.

Line graphs are another useful tool when dealing with data that is collected over time. The horizontal axis is used to create a timeline, while the data points that change over time are plotted along the vertical axis. You may already be quite familiar with these types of graphs. Remember, these are best used to represent *change over time*. We also see line graphs when examining supply and demand curves in business. In those cases, the continuous data is usually money.

In our example displaying the number of babies over a five-year period, the line graph for the data would look like:

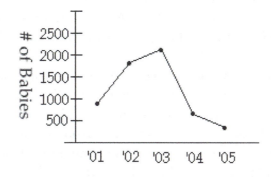

The **relative frequency** is the percentage of the data values that belong to a certain category. A **relative frequency table** is an extension of the frequency table that shows the relative frequency (or percentage) of each category that is listed. The relative frequency is calculated by dividing the number of data values in a given category by the total number of data values. Here, we have 20 total data values, so the relative frequency of the grade of "A" is: 4/20, or 0.20. We can read the information in this table as 20% of those who took the exam received As, 30% received Bs, etc. You may notice that, for any data set, the sum of the relative frequencies is equal to 1, or 100%.

Grade	Frequency	Relative Frequency
A	4	0.20 or 20%
B	6	0.30 or 30%
C	5	0.25 or 25%
D	1	0.05 or 5%
F	4	0.20 or 20%
Total:	20	

Pie charts, also called **circle graphs**, are used to show relative proportions (similar to relative frequencies). Again, you are probably very familiar with these types of graphs, but creating them may be new for you. With that in mind, we will use the exam data from above to create a pie chart.

The most important number to have when creating a pie chart is the total number of data values, which is the sum of the frequencies. In this case, we have 20 data values.

When making a pie chart, we need to decide how many degrees of the circle should be given to each category of data. Remember, a circle has 360 degrees in total, so that number will also be very important to us. Now, we have 20 total data values and 360 total degrees, so how do we divide up those 360°?

We begin by setting up a ratio, comparing the amount in a category to the total amount. For the grade of A, that ratio will be 4/20. A second ratio is also created using the degrees of a circle. If we let x be the number of degrees for the category, this ratio will be x/360.

Setting these two ratios equal to each other to make a proportion, we have 4/20 = x/360. Cross-multiplying gives us: 20x = 1,440. Dividing both sides of the equation by 20, we see x = 72. Thus, 72° will be allotted for the region of the pie chart corresponding to the grade of A, and it is customary to begin that measurement from the top of the circle, as if the left side of the wedge for the first category was pointing straight up from the center of the circle. Then, the right edge of that region (called a **sector**) is drawn at 72°, measured in a clockwise fashion from the starting point.

To determine the number of degrees given to the sectors for the B, C, D, and F categories, a similar process is done. However, although the B sector is to be given 108 degrees in our pie chart, we do not draw our line at 108° measured from the top of the circle. Instead, we start from the 72° line (where the sector for A grade ended), move clockwise 108°, and draw our line for B at the 180° mark (72° + 108° = 180°).

Finally, we need to make sure to label each sector in the pie chart. Continuing the above example, adding in the regions for C, D, and F, we have the following pie chart. Notice how the C grades make up 1/4 of the total (5/20 = 1/4), and, thus, the C sector is 1/4 of the pie chart.

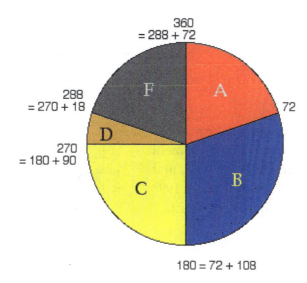

Section 3.4 Exercises

1. Average class sizes at a college are given in the table on the right. Create a bar graph that represents this data.

Class	Size
Math	34
English	27
History	22
Philosophy	17

2. The number of speeding tickets issued by the Mayberry police department during the months of May, June, and July of the years 1996, 1997, and 1998 are shown in the graph on the right.

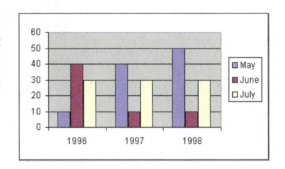

   a. How many tickets were issued in June of 1996?

   b. For the months shown on the graph, during which year were the most tickets issued?

   c. Over the three-year period, during which month were the most tickets issued?

3. A survey of college students that asked them about the mode of transportation that they used to travel to campus resulted in the following information: 120 students drove a car to school, 80 students rode a bike, 70 students took the bus, and 30 students walked to school. Find the number of degrees that should be given to each category, and create a pie chart that represents this data.

4. In a large physics class, 4 students earned As, 9 earned Bs, 14 earned Cs, 7 got Ds, and 6 students were issued Fs. Find the number of degrees that should be given to each category, and create a pie chart that represents this data.

5. A group of children purchased bags of jellybeans from a candy store. Later, they discovered 5 of the bags contained 12 jellybeans, 12 bags contained 13 jellybeans, 9 bags contained 14 jellybeans, and 4 bags had 15 beans. Use this data to create a relative frequency table and a bar chart. Round all decimals to the nearest hundredth.

6. The amount of time (in minutes) that it took a class of students to complete a short quiz was recorded in the table below. Use the data to create a histogram.

Minutes	# of Students
4	3
5	5
6	4
7	1
8	2
9	8
10	12

7. Use the following data to create a stem-and-leaf plot:

15, 23, 31, 17, 13, 20, 26, 23, 35, 11, 17

8. Use the following data to create a stem-and-leaf plot:

56, 84, 64, 91, 75, 59, 85, 58, 87, 66, 89, 73, 72, 66, 91, 81, 98, 63

9. Which type of graph would you use for the following set of data? Justify your choice.

Year	Population of Fictionland
1920	3,400
1940	3,900
1960	4,900
1980	6,300
2000	7,500

10. Which kind of graph is best for each of the following?

a. Displaying change over time.

b. Showing a comparison and including the raw data.

c. Showing a comparison that displays the relative proportions.

Answers to Section 3.4 Exercises

1. While you could arrange the classes in any order, the following is one way to create this graph and be correct:

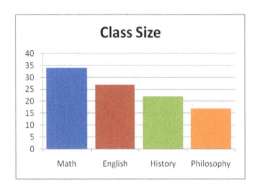

2. a. 40　　　　　　b. 1998　　　　c. May

3. While you could arrange the sections in any order within the graph, the number of degrees for each section are: Car = 144°; Bike = 96°; Bus = 84°; Walk = 36°

    One option for creating this graph and being correct is:

4. While you could arrange the sections in any order within the graph, the number of degrees for each section are: A = 36°; B = 81°; C = 126°; D = 63°; F = 54°

    One option for creating this graph and being correct is:

Grades

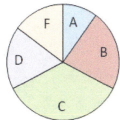

5. Frequency Table

# of Jelly Beans	# of Bags	Relative Frequency
12	5	0.17
13	12	0.40
14	9	0.30
15	4	0.13

30

Bar Chart

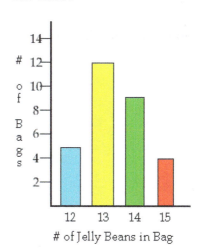

6. Histogram

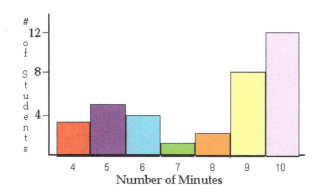

7.

Stem	Leaves
3	1 5
2	0 3 3 6
1	1 3 5 7 7

8.

Stem	Leaves
9	1 1 8
8	1 4 5 7 9
7	2 3 5
6	3 4 6 6
5	6 8 9

9. For this data, which shows change over time, the best choice would be to use a line graph or a histogram.

10. a. line graph, b. bar graph, c. pie chart

3.5 Is Anything Normal? (Standard Deviation & the Normal Curve)

Standard Deviation

Standard Deviation is a way to describe the amount of spread that is inherent in a data set. The standard deviation measures spread by finding the average distance of each data point from the mean. The more spread out the data, the higher the standard deviation will be.

To calculate the standard deviation of a data set, we will use the formula shown below. While this formula may appear to be intimidating at first, once you understand what all of the symbols represent, it really isn't a hard formula to use. You just have to be careful to remember the order of operations and execute them properly.

The symbol s, which is the lowercase Greek letter "sigma," represents the standard deviation.

The symbol m, which is the lowercase Greek letter "mu," (pronounced "m-you") represents the mean of the data set.

"n" is the total number of data points that are included in the data set.

Finally, we have these guys … : $x_1, x_2, x_3, \ldots x_n$

These are abbreviations for "x subscript 1," "x subscript 2," and so on, which are often pronounced using the shortened versions ... "x sub 1," "x sub 2," etc. These represent the individual data points, with x_1 being the first data point, x_2 being the second data point, and continuing until we get to x_n which is the last of the data points.

After all that buildup, here it is, the formula for calculating the standard deviation of a data set:

$$\sigma = \sqrt{\frac{\left(x_1 - \mu\right)^2 + \left(x_2 - \mu\right)^2 + ... + \left(x_n - \mu\right)^2}{n}}$$

To help take some of the mystery out of this formula, let's take a look at how it works. The key is to plug all of the proper numbers into the proper spots, and then, be careful as you simplify.

EXAMPLE 1: Calculate the standard deviation of the following data. Round your answer to the nearest hundredth.

45, 62, 73, 75, 90

Solution:

First, we must find the mean of our data set.

$\mu = (45 + 62 + 73 + 75 + 90)/(5) = 345/5 = 69$

Now, we will plug in the values of x_1, x_2, x_3, x_4, x_5, μ, and n into the formula, and simplify:

$$\sigma = \sqrt{\frac{\left(45 - 69\right)^2 + \left(62 - 69\right)^2 + \left(73 - 69\right)^2 + \left(75 - 69\right)^2 + \left(90 - 69\right)^2}{5}}$$

$$\sigma = \sqrt{\frac{\left(-24\right)^2 + \left(-7\right)^2 + \left(4\right)^2 + \left(6\right)^2 + \left(21\right)^2}{5}}$$

$$\sigma = \sqrt{\frac{576 + 49 + 16 + 36 + 441}{5}}$$

$$\sigma = \sqrt{\frac{1118}{5}}$$

$\sigma = \sqrt{223.6} = 14.95$ (to the hundredth)

So, the standard deviation for this data set is 14.95.

A data set with a smaller standard deviation would be less spread out than this one, and a data set with a higher standard deviation would be more spread out.

Right now, the concept of the standard deviation probably doesn't have a great deal of meaning for you, but as we move forward into studying the normal curve, we will see that the standard deviation plays a critical role in some generalizations we can make about certain types of data sets.

NOTE: Here, we have used the Greek letters mu (μ) and sigma (σ) to represent the mean and standard deviation of a data set. These letters, and the formula given in this section, are used when the data represents an entire population. If, instead, you are working with a data set that is a **sample** of the population, the symbol "$\overline{x}$" would be used to represent the mean, and "s" would be used to represent the standard deviation. Additionally, when working with a sample, the formula for standard deviation is slightly different. The standard deviation of a sample will not be covered in this text.

The Normal Curve

Below is a histogram of the heights of the students in a high school:

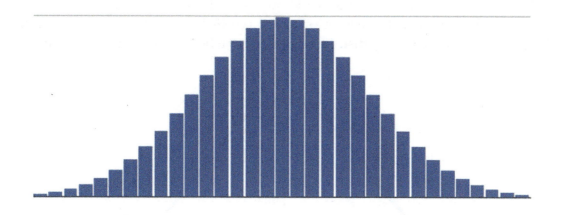

In an effort to streamline our data, we can draw in a curve that closely estimates the heights of the bars and, then, remove the bars from the diagram.

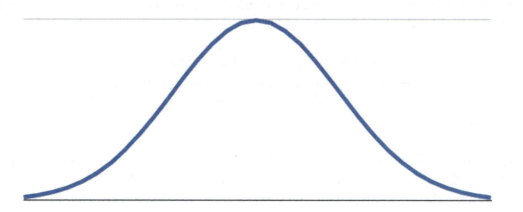

What remains in the figure above is a shape known as the **bell curve**, or a **normal curve**. Do you see why it is called a "bell" curve?

A data set like this one, in which the majority of the data points are close to the middle and the number of data points gradually tapers off as you move further away from the middle—with a very small number of data points in the extremes—is known as "normally distributed." When graphed, this type of data set creates a normal curve.

As it turns out, a great deal of real-world data actually fits into a shape like this. When graphed, normally distributed data form a symmetric, bell-shaped appearance. Also, the mean, median, and mode of a normally distributed data set will be right in the center of the curve.

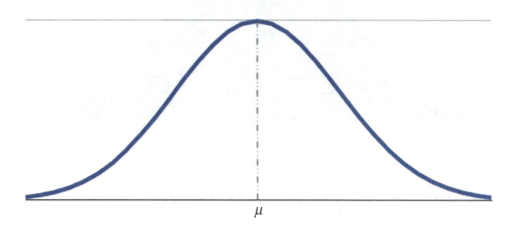

μ

Some examples of normally distributed data include:

- SAT/ACT scores (most are close to the mean, and as you move further away from the mean, the number of scores tapers off. A small number of scores are very low, a small number are very high)

- Birth weights of infants (most are close to the mean, a small number of babies are extremely light, a small number are extremely heavy)

- IQ scores (most people are close to the mean, a small number of people have very low IQs, a small number have very high IQs)

- Annual yields for a crop (most years are close to the mean level of production, a small number of years will result in very low production, and a small number of years will result in very large production)

- Life spans of batteries (most batteries will last an amount of time close to the mean, a small number will have very short life spans, a small number will have very long life spans)

Applications

Here we reach the point at which the concept of standard deviation becomes extremely useful, as it is closely tied to our study of the normal curve. On the normal curve shown below, the dotted line in the middle represents the mean of a normally distributed data set, and the other dotted lines represent multiples of the standard deviation.

That is, the "+1" visible under the curve indicates the location at which we are 1 standard deviation (SD) above the mean. The "−2" indicates 2 SD below the mean, and so on. Remember, the height of the curve corresponds the number of data points at a specific frequency (remember how we started with a histogram?).

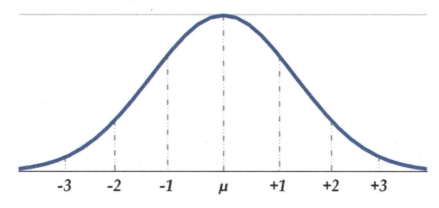

The 68–95–99.7 Rule

Normally distributed data is defined by its mean and standard deviation. Once we know that a data set is normally distributed, some very powerful observations can be made about the amount of data that is a given number of standard deviations from the mean. One of these observations is particularly useful, and it is known as the "68–95–99.7 Rule."

This rule states, for normally distributed data:

- 68% of the data points lie within 1 SD of the mean.

- 95% of the data points lie within 2 SD of the mean.

- 99.7% of the data points lie within 3 SD of the mean

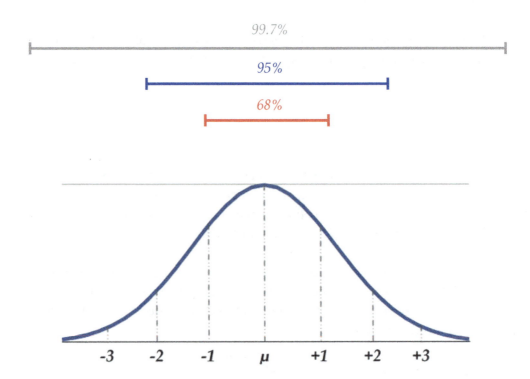

The dotted lines in the figure cut the area under the curve into several regions, and by using the 68–95–99.7 Rule, we can determine the amount of data within each particular region.

For example, since we know that 68% of the data is within 1 standard deviation of the mean, and the normal curve is always symmetric, we can conclude that 34% of the data is between the mean and 1 SD above the mean, as well as that 34% of the data is between the mean and

1 SD below the mean. Similarly, we can determine the amount of data that is within each specific region of the graph:

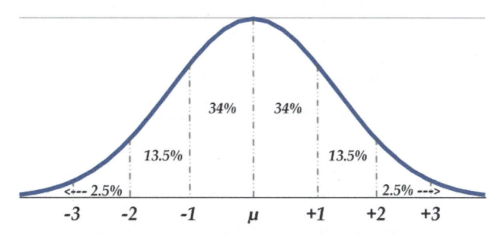

Note: The 2.5% labels shown in the diagram represent the data to the right (and left) of 2 SD away the mean. It does not represent the data that is in the region between 2 and 3 SD above the mean.

Disclaimer

Sometimes students fight the notion that we can be so sure about the location of all of these data points. Keep in mind, the 68–95–99.7 Rule, and other properties discussed in this section, only apply when we are *certain* the data is normally distributed. If the data is not normally distributed, then these rules do not apply. Remember, in this section, we will be working with data that do fit the normal curve. They are far more common than you might think.

EXAMPLE 2: The weights, in pounds, of the children in a third grade class were recorded, and the data was found to be normally distributed. The normal curve and the standard deviation indications for this data are shown below:

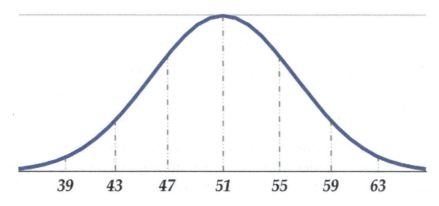

a. What is the mean weight of the children in this class?

b. What is the value of the standard deviation?

c. What percent of the weights are below 55 pounds?

d. What percent of the weights are above 43 pounds?

e. What percent of the weights are between 51 and 59 pounds?

Solutions:

a. The mean is in the middle of the curve: $\mu = 51$ pounds.

b. Since 55 represents one standard deviation above the mean, $s = 4$.

c. Since the mean is exactly in the middle, 50% of the data is below the mean. Adding this to the 34% of the data between 51 and 55 pounds, a total of 84% of the weights are below 55 pounds.

d. Adding the percentages within the necessary regions (13.5%+34%+50%), we find 97.5% of the children are heavier than 43 pounds.

e. Adding the percentages within the necessary regions (34%+13.5%), we find that 47.5% of the weights are between 51 and 59 pounds.

EXAMPLE 3: Given a set of normally distributed exam scores with a mean of 70 and a standard deviation of 8, find:

a. What percent of the scores were below 62?

b. What percent of the scores were above 70?

c. What percent of the scores were above 86?

d. What percent of the scores were between 54 and 78?

Solutions:

Even though we can answer the questions without an examination of the normal curve, it is a bit easier to do so. When drawing the curve, include the specific numbers for the mean and standard deviation.

1 SD above the mean is: $70 + 1 (8) = 70 + 8 = 78$.

2 SD above the mean is: $70 + 2 (8) = 70 + 16 = 86$.

3 SD above the mean is: $70 + 3 (8) = 70 + 24 = 94$.

Likewise, 1, 2, and 3 SD below the mean are 62, 54, and 46, respectively.

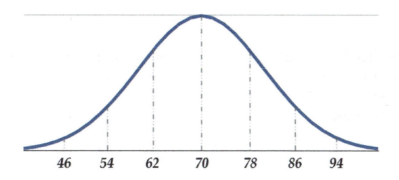

a. As we can see from the curve, 2.5%+13.5% of the data is in the region less than 62. Thus, 16% of the scores on the exam were below 62.

b. The mean score for this exam is 70. Since the data is normally distributed, 50% of the data is above this point.

c. 2.5% of the scores were above 86.

d. Adding the percentages within the necessary regions (13.5%+34%+34%), we find 81.5% of the scores are between 54 and 78.

Section 3.5 Exercises

1. Find the standard deviation for the following data. Round your answer to the nearest hundredth.

   {11, 16, 17, 20}

2. Find the standard deviation for the following data. Round your answer to the nearest hundredth.

   {30, 30, 31, 35, 39}

3. Considering the data from Exercises #1 and #2, which of the sets is more spread out? Why?

4. IQ scores are normally distributed with a mean of 100 and a standard deviation of 15.

   a. What percent of IQ scores are between 85 and 115?

   68%

   b. What percent of IQ scores are between 70 and 130?

   c. What percent of IQ scores are between 55 and 145?

d.   If the MENSA organization requires an IQ score higher than 130 for membership, what percent of the population would qualify for membership in MENSA?

5.   The heights (in inches) of a large number of students at a college were recorded, and the data was normally distributed. Use the normal curve shown below to answer the following questions:

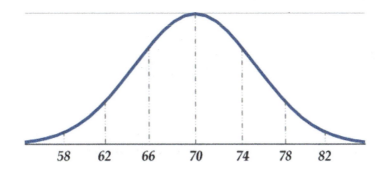

58   62   66   70   74   78   82

a.   What is the mean height for these students?

b.   What is the standard deviation?

c.   What percent of the students are taller than 62 inches?

d.   What percent of the students are shorter than 66 inches?

e.   What percent of the students are between 62 and 74 inches tall?

f.   What was the range for the middle 68% of student heights?

6.   The scores on a chemistry exam were normally distributed with a mean of 65 and a standard deviation of 9.

a.   What percent of the students scored above 65?

b.   What percent of the students scored above 56?

c.   What percent of the students scored below 47?

d.   What percent of the students scored between 65 and 74?

e.   What percent of the students scored between 56 and 83?

f.   If 1,000 students took the exam, and all students with grades between 74 and 83 earned a B, how many students earned a B?

g.   How high must a student score in order to be in the top 2.5% of scores?

7.   The life span for a certain brand of tires is normally distributed, with a mean of 35,000 miles and a standard deviation of 6,000 miles.

a. What percent of these tires will last more than 29,000 miles?

b. What percent of these tires will last fewer than 47,000 miles?

c. What percent of these tires will last between 23,000 and 35,000 miles?

d. What is the range for the middle 95% of the life spans of these tires?

8. The length of human pregnancies from conception to birth is normally distributed with a mean of 266 days and a standard deviation of 16 days.

a. What percent of human pregnancies last fewer than 266 days?

b. What percent of human pregnancies last between 266 and 298 days?

c. What percent of human pregnancies last more than 282 days?

9. A factory produces light bulbs that have a normally distributed life span, with a mean of 900 hours and a standard deviation of 150 hours. What is the percent chance that one of the light bulbs from this factory last at least as long as 750 hours?

10. A standardized test has a mean of 400 and a standard deviation of 50. A total of 2,000 students took this exam.

    a. How many students scored over 500?

    b. How many students scored at least 350?

11. The term "grading on a curve" is often heard. When teachers assign grades based on the notion that grades should be normally distributed, grades are assigned as follows:

    • Grades within 1 SD of the mean earn a C.

    • Grades between 1 SD above the mean and 2 SD above the mean earn a B.

    • Grades greater than 2 SD above the mean earn an A.

    • Grades between 1 SD below the mean and 2 SD below the mean earn a D.

    • Grades greater than 2 SD below the mean earn an F.

    If grading is done in this manner for a very large lecture class containing 400 students, how many of them would earn a C? How many would earn an A?

12. A student scored 73 on a standardized test that had a mean of 79 and a standard deviation of 3. Based on this score, what score would you expect for this student if he was to take another standardized test that has a mean of 95 and a standard deviation of 9?

13. Bonnie and Clyde attend different schools. Bonnie scored 84 on a math test that was normally distributed with a mean of 75 and a standard deviation of 9. Clyde scored 81 on a math test that was normally distributed with a mean of 70 and a standard deviation of 10. Which student did better on his or her exam?

Answers to Section 3.5 Exercises

1. 3.24

2. 3.52

3. Because it has a higher standard deviation, the data from Exercise #2 is more spread out.

4. a. 68%   b. 95%   c. 99.7%   d. 2.5%

5. a. 70 in   b. 4 in   c. 97.5%   d. 16%   e. 81.5%   f. 66–74 in

6. a. 50%   b. 84%   c. 2.5%   d. 34%   e. 81.5%   f. 135
   g. 83

7. a. 84%   b. 97.5%   c. 47.5%   d. 23,000 miles to 47,000 miles

8. a. 50%   b. 47.5%   c. 16%

9. 84%

10. a. 50 students   b. 1,680 students

11. 272 students would be given a C, and 10 students would get an A.

12. 77, which is a score 2 SD below the mean.

13. Clyde scored better. His score is more than 1 SD above the mean.

3.6 Lies, Damned Lies, and Statistics (Uses & Misuses of Statistics)

Former British Prime Minister Benjamin Disraeli once said "There are three kinds of lies: lies, damned lies, and statistics." In actuality, that line has been credited to Disraeli, William Shakespeare, Woody Allen, and others, but there is still a lot of truth to it. The statistical "lie" is often used to mislead the general public. People, in general, are usually not experts in understanding statistical data and, for that reason, tend to misinterpret such information when allowed to do so. To compound this "lie," we also tend to believe specific data when a seemingly knowledgeable person presents it to us.

Lie with Graphs and Pictures

The following graph indicates the time (in seconds) that it takes for two late model cars to accelerate from 0 to 60 miles per hour. Redraw this graph so it shows the entire scale from 0 to 11 seconds.

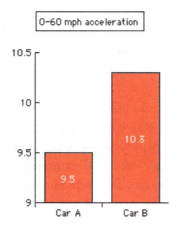

With a cursory glance, we may notice the bar for Car B is more than twice as high as the bar for Car A and assume this is because Car A accelerates over twice as fast as Car B. However, you will notice when you redraw the graph with the vertical axis ranging from 0 to 11 seconds, the difference between the cars appears to be much less significant.

When a graph-maker wants to make a difference look more significant (like in the graph shown above) it is a common practice to "truncate" the graph, and show a "zoomed-in" view of the vertical axis. To de-emphasize a difference, the graph-maker would "zoom out" and show the entire vertical axis (like we did by using a scale that ranged from 0 to 11). Manipulating pictures by deleting part of the vertical axis isn't really lying (the numbers *are* accurate), but it certainly can be a way of misleading the reader.

Lie by Not Telling the Whole Story

"Four out of five dentists recommended sugar-free gum for their patients who chew gum." We've all heard that claim (or something similar) before. Can you see some problems?

- Were only five dentists surveyed?

- How many times did they have to do this survey to get these results?

- Is it possible that only 1 out of 100 dentists recommended chewing gum at all?

Lie with Averages

Sometimes misinterpretations of statistics are the direct result of the use of the word "average." As we have seen, there are several different ways to compute averages. The two most common ones are the mean and median, but that is enough to create a "lie."

EXAMPLE 1: John has two job offers and his decision on which one to take rests solely on the salary. Company A states its average employee salary is $40,000, and Company B claims its average employee salary is $45,000. So, does this mean Company B is the better choice? Not necessarily. Examine the figures in the following table. Find the mean and median salary for each company.

Company A Salaries	Company B Salaries
$30,000	$30,000
$35,000	$30,000
$40,000	$30,000
$45,000	$30,000
$50,000	$105,000

Company A: mean = $40,000, median = $40,000

Company B: mean = $45,000, median = $30,000

Company B intentionally used the mean, instead of the median, when reporting its "average" salary. This isn't lying, but it is misleading to a potential employee. In this case, Company A is probably the better choice.

Here are a couple of statements that are commonly misinterpreted. What should the "proper" interpretation of each be?

1. On average, most auto accidents occur on Saturday night. This means that people do not drive carefully on Saturday night.

   Proper Interpretations: More people are on the road Saturday nights. Also, it is very likely that more people drink and drive on Saturday nights.

2. At East High School, half of the students score below average in mathematics. Therefore the school should receive more federal aid to raise standards.

   Proper Interpretation: If the reported "average" is the median, half the students will *always* be below average.

Lie through Omission

Caveat Emptor. Let the Buyer Beware. Warning… Below are some fictional advertisements, all of which could be true in some way, but all also "lie" by leaving out some relevant information. Can you spot the omissions?

1. Smoke beats Flame! In a recent taste test, an amazing 60 percent said Smoke cigarettes taste as good as or better than Flame cigarettes.

Caveat: In reality 36% of the people surveyed liked Smoke better, but 40% liked Flame better. When 24% said the two were equal in taste, the claim made becomes true. That's why the ad uses the words "as good as or better."

2. Hospitals recommend acetaminophen, the aspirin-free pain reliever in Ache-Free, more than any other pain reliever!

   Caveat: They casually neglected other national brands also contains acetaminophen, and hospitals recommended those other brands more than Ache-Free.

3. Ninety percent of college students say Votam Jeans are "in" on campus.

   Caveat: The ad cites a fall fashion survey conducted annually on 100 different college campuses. What was not mentioned was Votam Jeans were the only blue jeans listed in the survey. Other entries included T-shirts, 1960s-style clothing, overalls, and neon-colored clothing. So, anyone who wanted to choose any type of jeans had no choice but to pick Votam Jeans.

Section 3.6 Exercises

1. The number of years of experience of the kindergarten teachers at a school are as follows:

   Mr. Amazing, 2   Mrs. Super, 1

   Ms. Terrific, 10   Mrs. Dynamite, 2

   Mr. Great, 12   Ms. Fantastic, 3

   a. Find the mean, median, and mode for the number of years of experience for these teachers.

   b. If you were a parent, and you wanted to emphasize the fact that the teachers did not have very much experience, would you use the mean, median, or mode to support your argument?

   c. If you were the principal, and you were pointing out that your kindergarten teachers did have sufficient experience, would you use the mean, median, or mode to support your argument?

2. On a line graph that shows the change over time that occurs in the data, what can be done to make the change appear more drastic?

3. On a line graph that shows the change over time that occurs in the data, what can be done to make the change appear less drastic?

For Exercises #4 and #5, give a proper interpretation for each statement.

4. Eighty percent of all automobile accidents occur within 10 miles of the driver's home. Therefore, it is safer to take long trips.

5. The average depth of the pond is 3 feet, so it is safe to go wading.

6. The following two graphs show information regarding the price of fictional products called Florks. Both graphs use the same data for their creation.

   a. Describe ways in which each of these graphs could be used to mislead the reader.

   b. What could the manufacturer of Florks do with the graph that shows the actual price data, in order to improve public perception of the price increase over time?

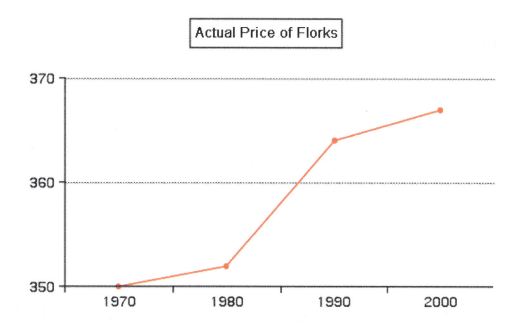

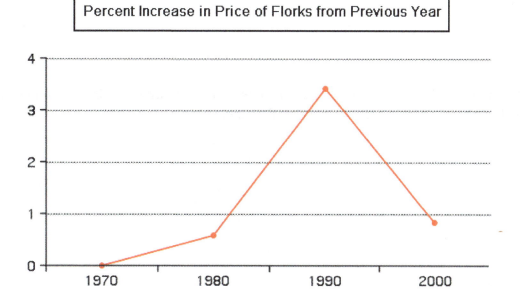

7. ABC Crackers claims, "Our crackers have 1/3 fewer calories." What is misleading about this claim?

8. What is misleading about the claim "Eating fish may help to reduce your cholesterol?"

9. What is improper about the survey question, "Are you going to vote for Candidate Jones, even though the latest survey shows he will lose the election?"

10. We can claim 71% of adults do not use sunscreen. Although 71% is a large percentage, explain what about that claim can be misleading.

Answers to Section 3.6 Exercises

1. a. mean = 5, median = 2.5, mode = 2
    b. The mode, by saying something like: "The most common number of years of experience is 2."
    c. The mean, by saying something like: "The average number of years of experience is 5."

2. Zoom in on the data by truncating the vertical axis in the graph.

3. Zoom out by using the entire vertical axis in the graph.

4. **Proper Interpretation:** Most accidents occur close to home, because when people drive, most trips are within 10 miles.

5. **Proper Interpretation:** The pond could be very shallow in some spots, but very deep in others.

6. a. The graph that shows the Actual Price of Florks could be used by *consumers* to show how the price has gone up dramatically over the years.
        The graph showing the percent increase in price could be used by the *manufacturer* to show the tremendous drop in the amount of increase from 1990 to 2000.
    b. The manufacturer could zoom out and show the entire vertical axis, resulting in a graph that looks like the following:

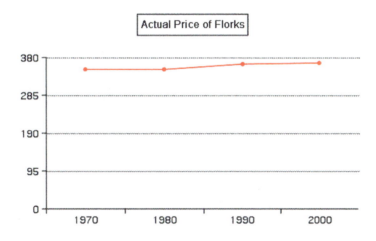

7. What are the crackers being compared to? If they had 1/3 fewer calories than a large piece of chocolate cake, that's not very good.

8. Notice the use of the word "may." You could just as easily claim, "Eating fish may not help to reduce your cholesterol."

9. Most people do not like to vote or do not admit voting for the losing candidate. This may lead people to say "No," thereby injecting bias into the results.

10. Not everyone spends a significant amount of time in the sun. Also, where was the survey conducted? If it were done in Seattle, the percent would be high. However, if it was done in Southern California, the results may indicate the majority of people do use sunscreen.

Credits

1. "Pie Chart," http://pixabay.com/en/pie-chart-diagram-statistics-parts-149727/. Copyright in the Public Domain.

2. Copyright © Wikimedia Foundation, Inc. (CC BY-SA 3.0) at http://en.wikipedia.org/wiki/Florence_Nightingale.

3. "Florence Nightingale," http://spartacus-educational.com/REnightingale.htm. Copyright in the Public Domain.

4. Copyright © Wikimedia Foundation, Inc. (CC BY-SA 3.0) at http://en.wikipedia.org/wiki/George_Gallup.

5. The Gallup Organization, "George Gallup," http://commons.wikimedia.org/wiki/File:George_Gallup.png. Copyright in the Public Domain.

6. Copyright © AleXXw (CC by 3.0 Austria) at http://commons.wikimedia.org/wiki/File:American_Football_EM_2014_-_AUT-DEU_-_165.JPG.

7. "Cat Face," http://pixabay.com/en/adorable-animal-background-218199/. Copyright in the Public Domain.

8. "Marathon Runners," http://pixabay.com/en/marathon-jogging-race-341299/. Copyright in the Public Domain.

9. "Baby and Stork," http://pixabay.com/en/baby-boy-boy-baby-cute-surprised-220312/. Copyright in the Public Domain.

10. "Bell," http://pixabay.com/en/bell-desk-table-leisure-booking-214297/. Copyright in the Public Domain.

11. "Broken Headlamp," http://pixabay.com/en/headlamp-accident-auto-blue-broken-2940/. Copyright in the Public Domain.

PROBABILITY

What's the difference between probability and statistics?

Simply put, to describe an entire set by looking at a few specific items randomly selected from the set, we use statistics. To attempt to predict the results of one particular event based on information that describes a set, we use probability.

Randomness is extremely important in the worlds of probability and statistics. If the systems under consideration were not random, they would be predictable. If they were predictable, then there would not be a need to study them like we do. Thus, to keep things truly random, unless otherwise stated, we will assume all events we study are random in nature and, unless specified otherwise, all out comes are equally likely. We further assume, in this book, all dice and coins are fair and balanced, all decks of cards are well shuffled, and all balls in urns have been thoroughly mixed.

4.1 On the Shoulders of Giants (Biographies & Historical References)

In correspondence with each other, Frenchmen Pierre de Fermat and Blaise Pascal laid the foundation for the theory of probability. This correspondence consisted of five letters and occurred in the summer of 1654. They considered the dice problem, which asks how many times one must throw a pair of six-sided dice before one expects a double six.

Even though Fermat and Pascal are credited with the foundations of probability theory, a rather quirky Italian doctor and mathematician named Girolamo Cardano was an expert in the probabilities involved in games of chance nearly 100 years earlier.

Blaise Pascal

Blaise Pascal (1623–1662) was a French mathematician, physicist, inventor, writer and Christian philosopher. His mother passed away when he was just three years old and his father, Étienne, decided that he alone would educate his children, for they all showed extraordinary intellectual ability, particularly his son Blaise. The young Pascal showed an amazing aptitude for mathematics and science.

In 1642, in an effort to ease his father's endless, exhausting calculations and recalculations of taxes owed and paid, Pascal, not yet 19, constructed a mechanical calculator capable of addition and subtraction, called "Pascal's Calculator" or the "Pascaline."

In 1654, he corresponded with Pierre de Fermat on the subject of gambling problems, and from that collaboration was born the mathematical theory of probabilities. The specific problem was that of two players who want to finish a game early and, given the current circumstances of the game, want to divide the stakes fairly, based on the chance each has of winning the game from that point. From this discussion, the notion of expected value was introduced. Pascal later used a probabilistic argument, which became known as Pascal's Wager, to justify belief in God and a virtuous life. It reads:

If God does not exist, one will lose nothing by believing in him, while if he does exist, one will lose everything by not believing. Thus, we are compelled to gamble.

The work done by Fermat and Pascal in the study of probabilities laid important groundwork for the formulation of calculus. His inventions include the hydraulic press (using hydraulic pressure to multiply force) and the syringe. Additionally, in his search for a perpetual motion machine, Pascal introduced a primitive form of roulette, and the roulette wheel.

After a religious experience in 1654, Pascal mostly gave up work in mathematics. His death came in 1662, just two months after his 39th birthday, with his last words being "May God never abandon me." An autopsy performed after his death revealed grave problems with his stomach and other organs of his abdomen, along with damage to his brain. Despite the autopsy, the cause of his poor health was never precisely determined, though speculation focuses on tuberculosis, stomach cancer, or a combination of the two.[1]

Pierre de Fermat

Pierre de Fermat (1601–1665) was a French lawyer and a mathematician who is given credit for early developments that led to the discovery of calculus. He made notable contributions in the fields of number theory, analytic geometry, probability, and optics, and, being fluent in six languages: French, Latin, Occitan, classical Greek, Italian, and Spanish, he was regularly sought out regarding the translation of foreign texts.

In 1623, he entered the University of Orléans and earned a bachelor's degree in civil law in 1626. Soon after, he began his first serious mathematical research, communicating most of his ideas in letters to friends. Although Fermat claimed to have proven all his theorems of arithmetic, few records of his proofs were actually included in these letters. This led many mathematicians to doubt several of his claims, especially given the difficulty of some of the problems and the limited mathematical methods available to Fermat, at the time. His famous "Last Theorem" was first discovered by his son in the margin of one of his father's books, and included the statement that "the margin was too small to include the proof."

Through their correspondence in 1654, Fermat and Blaise Pascal helped lay the fundamental groundwork for the theory of probability, and from this brief but productive collaboration, they are now regarded as joint founders of probability theory. Together with René Descartes, Fermat was one of the two leading mathematicians of the first half of the 17th century. According to Peter L. Bernstein, in his book *Against the Gods*, Fermat "was a mathematician of rare power."[2]

Girolamo Cardano

Girolamo Cardano (1501–1576), also known as Jerome Cardan, is one of the most interesting people in the history of mathematics. He was a mathematician, physician, astrologer, philosopher and gambler. He wrote more than 200 works on medicine, mathematics, physics, philosophy, religion, and music, and his gambling led him to formulate elementary rules in probability, making him one of the founders of the field.

He was born the illegitimate child of Fazio Cardano, a mathematically gifted lawyer, who was a friend of Leonardo da Vinci. After he earned a degree in medicine, Cardano managed to develop a considerable reputation as a physician and his services were highly valued. He was the first to describe typhoid fever, and in 1553 he cured the Scottish Archbishop of St Andrews of a disease that had left him speechless and was thought incurable—for which he was paid 1,400 gold crowns.

Today, he is best known for his achievements in algebra. Cardano was the first mathematician to make systematic use of numbers less than zero. He acknowledged the existence of what are now called imaginary numbers (although he did not understand their properties), and introduced the binomial coefficients and the binomial theorem.

Cardano was notoriously short of money and kept himself solvent by being an accomplished gambler and chess player. His book about games of chance, *Liber de Ludo Aleae* ("Book on Games of Chance"), was written around 1564, but not published until 1663, nearly 90 years after his death. The book contains the first systematic treatment of probability, as well as a section on effective cheating methods. He used the game of throwing dice to understand the basic concepts of probability, and defined odds as the ratio of favorable to unfavorable outcomes.

Cardano invented several mechanical devices including the combination lock and universal joints, which allow the transmission of rotary motion at various angles and is used in vehicles to this day. In a fit of rage, he cut off the ears of his youngest son, who had stolen money from him. Cardano himself was accused of heresy and had to spend several months in prison in 1570 because he had computed and published the horoscope of Jesus in 1554. Cardano's life came to a dramatic end. Years before, he had made an astrological prediction of the date of his own death. When the day arrived and he was still alive, he committed suicide to make his prediction come true.[3]

4.2 I'm Counting on You (Counting, Permutations, & Combinations)

Counting

Counting. Seems simple, right? We know how to count, or we wouldn't have made it into this math class. Well, yes and no. What we're going to examine are some different situations that require different types of counting.

Let's pretend we've gone into a restaurant, and the menu looks like this:

Appetizers	Main Dishes	Desserts
Breadsticks	Lasagna	Pie
Salad	Chicken	Ice Cream
	Meatloaf	

We are instructed by the waiter to choose exactly one item from each column to create a meal. How many different meals are possible?

One approach we can take in trying to answer this question is to make a list of all the possible meals. If we're going to do this, we should be as systematic as possible, so that we don't leave out any of the meals.

Going across the top row, the first meal that we could make would be: Breadsticks-Lasagna-Pie (we will abbreviate this as B-L-P). We could also have Breadsticks-Lasagna-Ice Cream (B-L-I). This takes care of every possibility that includes Breadsticks and Lasagna.

So far, we have:

B-L-P

B-L-I

Continuing, we will create every meal that has breadsticks as the appetizer:

B-C-P   B-M-P

B-C-I   B-M-I

That's it. That's every meal that has breadsticks as the appetizer. Now, we must consider salad as the appetizer:

S-L-P   S-C-P   S-M-P

S-L-I   S-C-I   S-M-I

Whew. Well, there they are; all 12 possible meals. If you're thinking, "there *must* be a better way," you're right. It is known as the **Fundamental Counting Principle**. The heart of the matter is as follows:

> Fundamental Counting Principle
> How many choices do you have for each option?
> Multiply those numbers together.

When we look at our menu from above, we had two choices of appetizer, three choices of main dish, and two choices of dessert. Multiplying these together gives us $2 \times 3 \times 2$, or 12 possible meals. Yep, it really is that easy.

EXAMPLE 1: A four-character code is required for entering a high-security room. This code is entered on a keypad that consists of the numbers one through nine and the letters A, B, and C. How many different four-character codes are possible?

We have four spots to fill with possible outcomes, and we are going to use the Counting Principle: Find the number of possibilities for each spot, and then, multiply those numbers together.

— — — —

The first spot in the code could be any of the 12 possible characters, so there are 12 possibilities for that spot:

12 — — —

The second spot could also be any of the 12 possible characters, so there are 12 possibilities for that spot, as well:

12 12 __ __

Similarly, the third and fourth spots also have 12 possibilities, so we have:

$12 \times 12 \times 12 \times 12 = 20{,}736$ different four-character codes that are possible

EXAMPLE 2: A four-character code is required for entering a high-security room. This code is entered on a keypad that consists of the numbers one through nine and the letters A, B, and C. *If repetition is not allowed*, how many different four-character codes are possible?

Here, since repetition is not allowed, once a character has been used it cannot be used again.

As in Example 1, we have four spots to fill with possible outcomes, and we are going to use the Counting Principle: Find the number of possibilities for each spot, and then multiply those numbers together.

__ __ __ __

The first spot in the code could be any of the 12 possible characters, so there are 12 possibilities for that spot:

12 __ __ __

Since one of the characters has been used in the first spot, we are left with 11 possibilities for the second spot:

12 11 __ __

Similarly, the third spot will have 10 possible characters and the fourth spot will have 9 possible characters, giving us:

$12 \times 11 \times 10 \times 9 = 11{,}880$ different four-character codes that are possible, when repetition is not allowed

Factorials

I have a class with six students in it, and their names are Red, Yellow, Green, Blue, Indigo, and Violet. We're going to line the students up in a single row for a class photograph. How many different photographs are possible? We will apply the Fundamental Counting Principle, and decide that for the photograph, we have to put a student in each of the six spots:

_____
_____

How many options do we have for the first spot? Well, no students have been used yet, so we have all six to pick from. We have six options for the first spot.

    6 _____

Now that we have placed a student (any one of them, it really doesn't matter which one) in the first spot, how many are left to choose from for the next spot? There would be five left to pick from, so we have five options for the second space. (I'm hoping you see the Fundamental Counting Principle at work here.)

    6   5 _____

Continuing on with the same logic, until we have only one student left to fill the last spot, our options look like this:

    6   5   4   3   2   1

We multiply these all together and wind up with $6 \times 5 \times 4 \times 3 \times 2 \times 1 = 720$ different ways. If this number seems too large, start listing them out, starting with RYGBIV, then RYGBVI, and so on. After you write down the first couple hundred permutations, you may get tired and accept the 720 as correct.

What if we were asked the same question about a deck of cards? How many different ways can you place the cards down in a straight line? Well, there are 52 cards, so that will be: $52 \times 51 \times 50 \times$ ... WAIT A MINUTE! You don't want to write that all the way down to 1, and doing all that multiplying on a calculator would be hard to do without making a mistake.

"There's got to be a better way to write this," you might say, and you would be correct. It is known as the **factorial** symbol.

The factorial symbol (commonly known as the exclamation point) gives us a shorthand way to write such problems down, and if you have this symbol on your calculator, it can save you a ton of work.

$$4! \text{ (pronounced "four factorial")} = 4 \times 3 \times 2 \times 1 = 24$$

Factorials get really large really fast.

$1! = 1$
$2! = 2 \times 1 = 2$
$3! = 3 \times 2 \times 1 = 6$
$4! = 4 \times 3 \times 2 \times 1 = 24$
$5! = 5 \times 4 \times 3 \times 2 \times 1 = 120$
$6! = 6 \times 5 \times 4 \times 3 \times 2 \times 1 = 720$
$7! = 7 \times 6 \times 5 \times 4 \times 3 \times 2 \times 1 = 5040$
$8! = 8 \times 7 \times 6 \times 5 \times 4 \times 3 \times 2 \times 1 = 40{,}320$
$9! = 9 \times 8 \times 7 \times 6 \times 5 \times 4 \times 3 \times 2 \times 1 = 362{,}880$
$10! = 10 \times 9 \times 8 \times 7 \times 6 \times 5 \times 4 \times 3 \times 2 \times 1 = 3{,}628{,}800$
$11! = 11 \times 10 \times 9 \times 8 \times 7 \times 6 \times 5 \times 4 \times 3 \times 2 \times 1 = 39{,}916{,}800$

$25! = 15{,}511{,}210{,}043{,}330{,}985{,}984{,}000{,}000$

Remember the question about the different arrangements for a standard deck of cards? Well, $52!$ is greater than 8.0658×10^{67}. That's a 68-digit number. The number one hundred trillion—100,000,000,000,000—only has 15 digits!

Our discussion of factorials would not be complete without a couple more pieces of the puzzle.

First, let's talk about $0!$. Although it is very tempting to think that $0! = 0$, that is incorrect. $0!$ is actually equal to 1. Some authors will make this claim "out of convenience" and others will claim it "as a definition." If you wish to accept $0! = 1$ without condition, so be it. If you would like a little proof, think of the following.

$$n! = n \times (n-1) \times (n-2) \times (n-3) \times \ldots \times 3 \times 2 \times 1$$

and

$$(n-1)! = (n-1) \times (n-2) \times (n-3) \times \ldots \times 3 \times 2 \times 1$$

Then, substituting $(n-1)!$ into the equivalent part of the first equation, we have

$$n! = n \times (n-1)!$$

Dividing both sides of the equation by n gives us

$$n!/n = (n-1)!$$

Now, if we let $n = 1$, the equation becomes

$$1!/1 = (1 - 1)!$$

And some minor simplification on both sides gives us

$1 = 0!$, which is the same as $0! = 1$.

Finally, since we have covered factorials for 0 and the positive integers (i.e., the Whole Numbers), one might be tempted to ask about factorials for fractions, decimals, or even negative numbers. Well, the good news is, for our purposes, we will only work with factorials involving Whole Numbers.

Permutations

A **permutation** is an arrangement of objects that are placed in a distinct order. Just like in our class photo example from earlier in the section, if the order changes, then we have a new permutation. We worked with this idea in our photograph example, and that helped explain factorials. The photo RGYBIV is different from the photo BIVGYR. The letters are the same, but the order has changed, creating a different photograph. Since changing the order makes the photo different, we are dealing with a permutation. **For a permutation, the order matters**.

As long as individual items are not being duplicated, the formula for finding the number of permutations is as follows, where n is the total number of objects, and r is the number of objects being used. I know, "being used" doesn't make sense yet, but it will as we continue to practice.

$$_nP_r = \frac{n!}{(n-r)!}$$

Although many people resort to memorizing that formula, it makes a bit more sense if we take a few moments to understand it.

Permutations are all the possible ways of doing something. Take another look at our photo example. What if we wanted only four of the six students in the picture? How many ways could it be done? Formally, this is called a "Permutation of six things, taken four at a time," or $_6P_4$.

From the Fundamental Counting Principle, we have 6 choices for the first person, 5 for the second, 4 for the third, and 3 choices for the last student. That makes $6 \times 5 \times 4 \times 3 = 360$ different photos.

Alternatively, if we begin by considering all the possible arrangements with six students, we have 6!. Then, because we have two students left out of the photo, we would need to cancel the possible arrangements of those students. With numbers, this looks like 6!/2!. Thus, the 2×1 would cancel, leaving $6 \times 5 \times 4 \times 3$ in the numerator.

Relating that to the formula, $_nP_r = \dfrac{n!}{(n-r)!}$, the $n!$ in the numerator refers to all possible arrangements. For the denominator, if we have n items and only want to count r of them, then there are $(n - r)$ of them we don't want to count. The $(n - r)!$ in the denominator refers to the arrangements that are not counted.

Now, going back to the class picture example,

$_6P_4 = 6!/(6-4)! = 6!/2! = (6 \times 5 \times 4 \times 3 \times 2 \times 1)/(2 \times 1) = 6 \times 5 \times 4 \times 3 = 360$

There are 360 different ways to take a photo using four of the six different students.

Combinations

A **combination** is a collection of distinct objects in which the order makes no difference. Distinguishing between combinations and permutations is often challenging, but remember, for combinations, order does not matter.

Let's suppose, for example, of six students in a class, four of them will be chosen to go on a field trip. Here, we are making a collection of four students, and the order in which they are chosen makes no difference. Each student is either chosen, or not, but it doesn't matter which one is chosen first. While RGBY would make a different *photo* than GBRY (which indicates a **permutation**), having RGBY go on the field trip is exactly the same as having GBRY go on the field trip (which indicates a **combination**). The distinction between the two can be difficult. Try to decide if the order would make a difference. If not, then we have a combination.

In terms of combinations, RGBY is the same as GBRY. In fact, any group of those four students is the same. They are all redundant, because order does not matter. Furthermore, in this situation, there are 4! = 24 redundancies.

Finding the number of combinations starts with finding the number of permutations, and then involves dividing by the number of redundancies.

Just like in the formula for permutations, as long as individual items are not reused, the formula for determining the number of combinations is as follows, where n is the total number of objects, and r is the number of objects being used.

$$_nC_r = \frac{n!}{(n-r)!\,r!}$$

Think about that formula for a minute. Just like with the formula for permutations, $n!$ represents the total number of all possible arrangements and $(n - r)!$ has us cancelling out the unused arrangements. Then, $r!$ has us cancelling out the redundancies. It may also be helpful to remember, since we cancel out the redundancies, the number of combinations for a situation can never be more than the number of permutations for the same situation.

Going back to our example, let's say we want to pick four out of six students to take on the field trip. Here, we have six students, but only four of them will be chosen. Formally, this is called a "Combination of six things, taken four at a time," or $_6C_4$.

Using the formula, we get,

$$_6C_4 = \frac{6!}{(6-4)!\,4!} = \frac{6!}{2!\,4!} = \frac{6\cdot5\cdot4\cdot3\cdot2\cdot1}{(2\cdot1)\cdot(4\cdot3\cdot2\cdot1)} = \frac{6\cdot5}{2\cdot1} = \frac{30}{2}$$

There are 15 different ways to pick four of the six students to take on the field trip.

A key thing to notice in working the formula for combinations is that the factorials in the denominator are not multiplied together to form a single factorial. Combining the 4! and the 2! to make 8! would be incorrect. Don't do that.

Notice that the numbers used in the combination example are the same as those used in the example for the permutation, but the answers are *very* different.

Finally, in regards to the notation, we indicate permutations and combinations by sub-scripting the n and the r in $_nP_r$ and $_nC_r$. As an alternative, we could write nPr and nCr without subscripting the n and r. Also, in some texts, you may see these written as P(n, r) and C(n, r), respectively. In all cases, though, the P and C are capital letters, and the n and r are lowercase letters that are usually replaced with numbers.

One More Hint

Often, the hardest part about working with permutations and combinations is determining which one to use—based on the description of the situation. Sometimes, certain key words can help us decide if we have a permutation or a combination—not always, but usually ...

Term	Indicates
arrangement, order	Permutation
choose, select, pick	Combination

In general, *always* ask yourself, "Does order matter?" If so, we have a permutation. If not, it is a combination.

Also think about the logistics of the situation. If we have a box of 50 photographs, and we want to pick 15 of them to put into an album, we are working with a combination, because it does not matter in what order we choose them—we just have to pick 15 of them. If, however, we are actually putting the photos in the photo album, the order makes a difference. Putting a picture of your mother on the first page as opposed to the tenth page makes a difference. If the order matters, we have a permutation.

Using Your Calculator

If you have a scientific calculator, you probably have a factorial key on it. Type in the number, and look for a button labeled "x!" or "n!." On some calculators, once you hit the factorial button, the computation will happen right away—without even having to hit the = key. On other calculators, you will have to hit =.

If your calculator can perform permutations and combinations, the keys will be labeled "nPr" and "nCr," respectively. Usage of these functions varies from calculator to calculator. In most cases, you type in the value for n, hit the nPr (or nCr) key, then type in the value for r, and hit the = key. If this sequence of entries does not work on your calculator, consult your calculator's owner's manual.

Do keep in mind; even if your calculator performs factorials, permutations, and combinations, you still need to be able to recognize which of those functions applies to the given situation. Remember, the calculator is only as accurate as the person using it.

Section 4.2 Exercises

1. A fraternity is to elect a president and a treasurer from the group of 40 members. How many ways can those two officers be elected?

2. Sandra has nine shirts and five pairs of pants. Assuming that everything matches, how many different outfits can she make?

3. Area codes are made up of three-digit numbers.
   a. If all numbers can be used for each digit, how many area codes are possible?

   b. How many area codes would be possible if repetition is not allowed?

4. A school gymnasium has six different doors. How many ways can a person enter the gymnasium and leave through a different door?

5. A license plate consists of three letters followed by three numbers.

   a. How many of these license plates are possible?

   b. How many plates are possible if repetition is not allowed?

Evaluate:

6. 4!

7. 7!

8. 11!

9. $\dfrac{9!}{5!}$

10. $_7P_3$

11. $_5P_5$

12. $_5P_1$

13. $_9P_5$

14. You have eight books to place on a shelf, but you only have room for five of them. How many different ways can you arrange books on this shelf?

15. Sam has five favorite football teams, and every week, he puts their flags on his flagpole in random order.
   a. In a given week, how many different ways can he arrange the flags?

   b. If he only has room for three of the flags on his flagpole, how many different ways can they be arranged?

16. A baseball batting order is made up of nine players. How many different batting orders are possible?

17. Ten runners are in a race. How many different ways can they finish in 1st, 2nd, and 3rd place?

Evaluate:

18. $_7C_4$ 19. $_8C_3$ 20. $_5C_5$ 21. $_6C_0$

22. Of seven students in a class, two will be chosen to go on a field trip. How many different ways can they be selected?

23. Of the 15 players at an awards dinner, 3 of them will be given identical trophies. How many different ways can the trophies be given out?

24. A teacher chooses 5 of her 12 students to help clean the room after school. In how many ways can the students be chosen?

25. A student must answer five of the nine essay questions that are on an exam. How many ways can the student select five questions?

26. What is the difference between a permutation and a combination? Give an example of each.

Answers to Section 4.2 Exercises

1. 1560

2. 45

3. a. 1000, b. 720

4. 30

5. a. 17,576,000, b. 11,232,000

6. 24

7. 5040

8. 39,916,800

9. 3024

10. 210

11. 120

12. 5

13. 15,120

14. 6720

15. a. 120, b. 60

16. 362,880

17. 720

18. 35

19. 56

20. 1

21. 1

22. 21

23. 455

24. 792

25. 126

26. When the order of the items matters, you have a permutation. When the order does not matter, you have a combination.

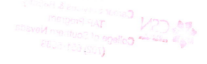

4.3 What Happens in Vegas (Simple Probability & Odds)

The Basics of Probability

A **sample space** is the list of everything that could possibly happen during an experiment, such as flipping a coin. It is common to list the sample space as a set, and the sample space for flipping a coin would be relatively short: {heads, tails}, or to abbreviate: {H, T}. Each result of the experiment is called an **outcome**, and outcomes can be listed individually or collectively as **events**.

Thus, if we are rolling a single six-sided die and are interested in the number of dots on the top side, our sample space, which contains six different outcomes, would be: {1, 2, 3, 4, 5, 6}.

An example of an event would be rolling a three, which would correspond to the set: {3}. Another example of an event would be rolling an even number, which would correspond to the set: {2, 4, 6}.

There are two different kinds of probability

More formal definitions are out there, but what it boils down to is two different situations: Making predictions on what *should happen*, and making predictions based on what *has already happened*.

Theoretical Probability

This is what *should happen*. Flipping a coin, we would expect heads to occur half of the time. If we flipped a coin 100 times, we would expect to get about 50 heads. **Theoretical probability** (or just **probability**) is calculated as the likelihood of obtaining a specific event.

Relative Frequency

This is what *has already happened*. In different publications, we often see this referred to as "empirical probability," "estimated probability," or even "experimental probability." To avoid confusing it with theoretical probability, we will avoid the word "probability" and refer to it as **relative frequency** (or just **frequency**). To discover the relative frequency of an event, we must conduct an experiment or a study and use those results to predict future results. If we flipped a coin 100 times and got 97 heads, we might suspect that this coin wasn't fair, and we could use those results to say we are much more likely to get a head on the next flip.

Disclaimers

If it is impossible to predict all possible outcomes or the outcomes are not equally likely, then probabilities can only be determined from the result of an experiment and, hence, are actually frequencies. For example: In baseball, it is impossible to determine, before he goes to bat, whether or not a player will get a hit (the outcomes are not equally likely), but based on all of his previous at bats, a fair prediction can be made. The player's batting average is, essentially, a relative frequency being used to make a prediction. Thus, based on previous at bats, a player with a batting average of .315 has a 31.5% chance of getting a hit on his next at bat.

Because some entities use the term "probability" in describing both theoretical probability and relative frequency, it can be, at times, difficult to distinguish between the two. In an ideal case, the two will be equal, and it would not matter. And, after an experiment, if the two are very different, it is usually an indication of an unusual occurrence.

In general, the problems that we will be working on will be asking you to find a theoretical probability, and unless otherwise noted, we also need to assume each outcome has an equal chance of occurring.

Next, if you remember only one thing about probability, remember this:

$$probability = \frac{\# \ successes}{\# \ outcomes}$$

Probability = (# of Successes)/(# of Outcomes)

A key thing to note here is that the term "success" is not necessarily something that corresponds to winning. We have a "success" if the event for which we are interested in finding the probability actually occurs.

EXAMPLE 1: A single six-sided die is to be rolled. What is the probability of rolling a four? Alternatively, using shorter notation, find P(4). Also, what is the probability of rolling an even number, or P(even).

In both cases, the sample space is {1, 2, 3, 4, 5, 6}. Since there are exactly six possible outcomes, that will be the denominator of our answer.

For P(4), the number of successes (notice, we don't *win* anything), or fours, on the dice is one. That will be the numerator of our answer. So, the probability of rolling a four is P(4) = 1/6.

For P(even), we succeed when the roll yields a two, four, or six. So, the probability of rolling an even number is P(even) = 3/6.

This answer can be reduced to 1/2, but that is not necessary when working probability problems. The reason for leaving the fraction unreduced is because an answer can provide us information about the raw number of successes and outcomes for the situation we are considering. With P(even) = 3/6, we can tell there were three successes out of six attempts. If the fraction were reduced to P(even) = 1/2, we would not be able to tell how many successes or outcomes were there originally.

We could also represent this answer 0.50 or even 50%. These are not incorrect, but it is more common to write probabilities as fractions, instead of decimals or percents.

EXAMPLE 2: If we flip a coin and roll a die, find the probability of getting heads on the coin and a one on the die.

With two possible events for the coin flip and six events for the die, the sample space consists of twelve unique outcomes: {H1, H2, H3, H4, H5, H6, T1, T2, T3, T4, T5, T6}.

Exactly one of those outcomes is H1. So, P(H1) = 1/12.

Rolling a Pair of Dice

It is very common to examine problems that involve rolling a pair of dice and taking the sum of the faces that are showing. There are 6 different outcomes on each die, and we can use the counting principle to find that there will be 36 (6 × 6) different outcomes in this situation.

The following table is a way to list all 36 of those outcomes, with the sum of the numbers showing on top of the dice being at the top of each column in the table.

Note: In each pair of numbers, the first number represents the result of rolling the first die, and the second number represents the result of rolling the second die.

2	3	4	5	6	7	8	9	10	11	12
(1,1)	(1,2)	(1,3)	(1,4)	(1,5)	(1,6)	(2,6)	(3,6)	(4,6)	(5,6)	(6,6)
	(2,1)	(2,2)	(2,3)	(2,4)	(2,5)	(3,5)	(4,5)	(5,5)	(6,5)	
		(3,1)	(3,2)	(3,3)	(3,4)	(4,4)	(5,4)	(6,4)		
			(4,1)	(4,2)	(4,3)	(5,3)	(6,3)			
				(5,1)	(5,2)	(6,2)				
					(6,1)					

EXAMPLE 3: Two dice are rolled and the sum of the faces is obtained. Find the probability that the sum is nine.

The number of outcomes when rolling a pair of dice is 36. That will be the denominator of our answer.

The number of successes, or ways to roll a nine, in our list of outcomes is four. That will be the numerator of our answer.

So, the probability of getting a sum of nine is P(9) = 4/36. This can be reduced to 1/9, but this reduction is not necessary.

Deck of Cards

Problems involving a standard deck of cards are very common. There are 52 cards in a standard deck, and they are divided into 4 suits, as follows:

Black Cards

- ♠ — Spades: {Ace, 2, 3, 4, 5, 6, 7, 8, 9, 10, jack, queen, king} ... 13 cards

- ♣ — Clubs: {Ace, 2, 3, 4, 5, 6, 7, 8, 9, 10, jack, queen, king} ... 13 cards

Red Cards

- ♥ — Hearts: {Ace, 2, 3, 4, 5, 6, 7, 8, 9, 10, jack, queen, king} ... 13 cards

- ♦ — Diamonds: {Ace, 2, 3, 4, 5, 6, 7, 8, 9, 10, jack, queen, king} ... 13 cards

Since the jack, queen, and king of each suit have pictures of people on them, they are often referred to as "face cards" or "picture cards."

EXAMPLE 4: In selecting one card from a standard deck, find the probability of selecting a diamond.

The number of outcomes when picking a card is 52. That will be the denominator of our answer. The number of successes, or diamonds, in the deck is 13. That will be the numerator of our answer.

So, the probability of picking a diamond is P(♦) = 13/52. As before, this can be reduced to 1/4, but this is not necessary.

EXAMPLE 5: In selecting one card from a standard deck, find the probability of selecting a five or a heart.

The new idea here is the presence of the word "or," which tells us that *both* fives and hearts are now to be considered successes.

The number of outcomes when picking a card is 52. That will be the denominator of our answer. The number of fives in the deck is four, but these are not our only successes. When figuring out our numerator, we also must consider the hearts. There are 13 of them, but we have already counted the five of hearts, so there are 12 hearts still to be included. Adding all our successes together, we have 16 of them, and that will be the numerator of our answer.

So, the probability of picking a five or a heart is P(5 or ♥) = 16/52. Once again, this can be reduced to 4/13, but this is not necessary.

EXAMPLE 6: In selecting one card from a standard deck, find the probability of selecting an ace and a spade.

The new idea here is the presence of the word "and," which tells us that a card must be an ace *and* a spade to be considered a success.

The number of outcomes when picking a card is 52. That will be the denominator of our answer. The number of successes, or cards that meet *both* conditions, is 1. That will be the numerator of our answer.

So, the probability of picking an ace and a spade, or P(ace and ♠) = 1/52.

A Couple of Important Observations

On a roll of a single six-sided die, find P(20).

It can't happen, right? There is no way to roll a 20 on a single six-sided die. So, let's see what that would look like as a probability.

- The number of outcomes is six, and the number of successes is zero, so:

- P(20) = 0/6 = 0

Something that can never happen has a probability of 0.

On a roll of a single die, find P(number less than 10).

Well, all of the numbers are less than 10, so this will happen for sure.

- The number of outcomes is six, and the number of successes is six, so:

- **P(number < 10) = 6 / 6 = 1**

> Something that will definitely happen has a probability of **1**.

The two previous examples lead us to a very important conclusion:

> All probabilities are between zero and one, inclusive.

The good news here is that we can use this fact to check our answers. The bad news is, if we are asked to find a probability and give an answer that does not fall between 0 and 1, it shows we do not have much of an understanding of probability, at all.

Odds

The "odds" a of particular event are a relatively simple idea, but be aware, they can occur in two ways: odds in favor and odds against. The **odds in favor** of an event are computed as the ratio of the *chance of successes* to *chance of failures*. The **odds against** a particular event are computed as the ratio of *chance of failures* to *chance of successes*. Thus, if we know the odds against an event are 10 to 1, in 11 identical attempts, we should expect 10 failures and 1 success of the event in question. We *must* be sure of the situation with which we are being presented! For example, in almost all racetracks, casinos, and sportsbooks, the stated "odds" are always given as odds against. Also, odds statements can be expressed using the word "to," as a ratio using a colon to separate the two numbers, or even a fraction. To keep things simple, we will state our odds statements as a ratio using a colon to separate the two numbers.

> Express odds statements as a ratio using a colon to separate the two numbers.

Be aware; the "odds" stated in racetracks, casinos and sportsbooks are not actually true odds statements. In reality, they are just **pay schedules**. That is, if a casino lists the odds of a team winning a game as 20/1 (read "20 to 1"), that means, if the team wins, they will pay $20 for every $1 bet.

Furthermore, the pay schedules offered in sportsbooks are not even their projected odds of a team winning, or even the representation of the results for a specific event. They create these "odds" statements based on their knowledge, experience, and even inside information, but these statements are actually an attempt to control the flow of money bet on a specific game or event. If a sportsbook sees a large amount of money being bet on one team in a game, they will adjust the payout so future wagers on that team do not pay as much.

Also, although they can predict the likelihood of one team winning, they cannot be certain. For this reason, they would prefer to have an equal amount of money bet on both teams. Then, they make their money by taking a small percent of the total amount that has been wagered.

For table games like Blackjack, the so-called odds statements given to gamblers are not true odds, either. Some casinos pay 2 to 1 for a Blackjack and others may pay 6 to 5 with, otherwise identical rules. These payouts are always a bit less than they would be if they were based on the true odds in the game. That way, the casino will make money.

For our purposes, however, we are going to assume the statements are reflective of the true odds of the event.

EXAMPLE 7: The probability of an event is 21/38. What are the odds in favor of the event? What are the odds against?

Since the p(success) = 21/38, there would 21 successes out of 38 outcomes. This means there would be 38 − 21 = 17 failures.

Then, the odds in favor of the event are 21:17, and the odds against the event would be 17:21.

EXAMPLE 8: If the odds against an event are 3:11, what is the probability of success for the event in question?

From the odds statement, there are 3 failures and 11 successes. This means there must be 14 (3 + 11) outcomes.

Thus, p(success) is: 11/14.

Section 4.3 Exercises

For Exercises #1 through #11, a dime is flipped and a single die is rolled.

1. Use the counting principle to determine the total number of outcomes.

2. List the sample space of all possible outcomes.

3. Find the probability of getting heads and a three.

4. Find the probability of getting tails and a seven.

5. Find the probability of getting heads and an odd number.

6. Find the probability of getting heads and a number greater than six.

7. Find the probability of getting tails or a seven.

8. Find the probability of getting tails or a number less than nine.

9. Find the odds against obtaining a head.

10. Find the odds against obtaining a six.

11. Find the odds in favor of obtaining an even number.

For Exercises #12 through #16, a quarter is flipped and a penny is flipped.

12. Use the counting principle to determine the total number of outcomes.

13. List the sample space of all possible outcomes.

14. Find the probability of getting tails on the quarter and tails on the penny.

15. Find the probability of getting exactly one tails.

16. Find the probability of getting no tails.

For Exercises #17 through #21, the letters of the word TOWEL are written on slips of paper and placed into a hat. Two of these letters will be pulled out of the hat, without replacement, one after the other.

17. Use the counting principle to determine the total number of outcomes.

18. List the sample space of all possible outcomes.

19. Find the probability that the first letter is a vowel.

20. Find the probability that both letters are vowels.

21. Find the probability that neither letter is a vowel.

For Exercises #22 through #26, a pair of dice is rolled, and the sum of the faces is obtained.

22. Find the probability that the sum is six.

23. Find the probability that the sum is odd.

24. Find the probability that the sum is divisible by three.

25. Find the odds against the sum being 11.

26. Find the odds in favor of the sum being five.

For Exercises #27 through #31, data regarding the transportation of children to school is given in the table on the right.

Use the information in the table to find:

	Bus	Car
Boys	9	5
Girls	3	8

27. The probability that a student chosen at random is a girl.

28. The probability that a student chosen at random rides the bus.

29. The probability that a student chosen at random rides in a car.

30. The probability that a student chosen at random is a girl who rides the bus.

31. The probability that a student chosen at random is a boy who rides in a car.

For Exercises #32 through #36, a bag contains 24 jellybeans, with the colors as follows:

Red = 5, Blue = 3, Orange = 7, Green = 4, Yellow = 2, Purple = 3

32. Find the probability that a jellybean selected at random is orange.

33. Find the probability that a jellybean selected at random is green or yellow.

34. Find the probability that a jellybean selected at random is neither red nor purple.

35. Find the probability that a jellybean selected at random is anything but yellow.

36. Find the probability that a jellybean selected at random is black.

For Exercises #37 through #45, a single card is drawn from a standard deck of 52 cards

37. Find the probability that the card is the queen of hearts.

38. Find the probability that the card is a six.

39. Find the probability that the card is a black card.

40. Find the probability that the card is a red nine.

41. Find the probability that the card is an ace or a heart.

42. Find the probability that the card is an ace and a heart.

43. Find the probability that the card is a face card.

44. Find the odds against drawing a king.

45. Find the odds in favor of drawing a queen or a diamond.

46. The odds against an event are given below. Find the probability of each event.

    a. 5:2   b. 3:7

47. What is the difference between theoretical probability and the relative frequency of an event?

Answers to Section 4.3 Exercises

Fractional answers that are reduced are acceptable, but the fractional answers given here have not been reduced.

1. 12

2. {H1, H2, H3, H4, H5, H6, T1, T2, T3, T4, T5, T6}

3. 1/12

4. 0

5. 3/12

6. 0

7. 6/12

8. 1

9. 6:6

10. 10:2

11. 6:6

12. 4.

13. {HH, HT, TH, TT}

14. 1/4

15. 2/4

16. 1/4

17. 20

18. {TO, TW, TE, TL, OT, OW, OE, OL, WT, WO, WE, WL, ET, EO, EW, EL, LT, LO, LW, LE}

19. 8/20

20. 2/20

21. 6/20

22. 5/36

23. 18/36

24. 12/36

25. 34:2

26. 4:32

27. 11/25

28. 12/25

29. 13/25

30. 3/25

31. 5/25

32. 7/24

33. 6/24

34. 16/24

35. 22/24

36. 0

37. 1/52

38. 4/52

39. 26/52

40. 2/52

41. 16/52

42. 1/52

43. 12/52

44. 48:4

45. 16:36

46a. 2/7, b. 7/10

47. Experimental probability is the result of performing trials, and theoretical probability is calculated.

4.4 Vegas Revisited (Compound Probability & Tree Diagrams)

Compound Probability

Events that occur one after the other, or in conjunction, are called **successive events**. When a series of successive events is necessary to arrive at an outcome, the individual probabilities are multiplied together to obtain the probability of that outcome. This is called **compound probability**, and sounds a lot more confusing than it really is. Let's take a look.

EXAMPLE 1: An urn contains six red, three white, and four blue balls. A ball is selected and then replaced, and then another ball is selected. Find the probability that the first ball is white and the second ball is blue.

First, we find the individual probabilities:

P (first ball is white) = 3/13

P (second ball is blue) = 4/13

Since we need *both* of these events to happen to have a success, we multiply the probabilities together:

P (W, then B) = (3/13) x (4/13) = 12/169

EXAMPLE 2: Using the same urn from above, but this time once the first ball is selected it is *not* placed back into the urn. Find the probability that the first ball is white and the second ball is blue. In this example, we are drawing *without* replacement.

First, we find the individual probabilities:

P (first ball is white) = 3/13

P (second ball is blue) = 4/12 (notice, now there are only 12 balls left in the urn)

Since we need *both* of these events to happen to have a success, we multiply the probabilities together:

P (W, then B) = (3/13) x (4/12) = 12/156

We could certainly reduce the answer to 1/13 if we want to, but it is not necessary.

EXAMPLE 3: Ten of the 25 players on the Yankees make over a million dollars per year. Eleven of the 12 players on the Lakers make over a million dollars per year. If one player is selected at random from each of the teams, what is the probability that they will *both* make over a million dollars a year?

We find the two individual probabilities:

P (Yankee makes over a million per year) = 10/25

P (Laker makes over a million per year) = 11/12

Since we need them *both* to make over a million per year, we need both events to happen, so we multiply the probabilities together:

P (both players chosen make over a million/year) = (10/25) × (11/12) = 110/300, which means they will *both* make over a million dollars per year about 1/3 of the times we try this activity.

The vast majority of our successive events problems will follow the setup of those listed above, in which we are able to simply multiply the probabilities together to arrive at the result. When the problems become more difficult, a tree diagram can be used to illustrate the situation.

Tree Diagrams

A **tree diagram** is a graphical representation of all the possible outcomes of an activity. It literally looks like a tree turned onto its side or upside-down. To create a tree diagram, start with a single point. Then, draw and label a branch for each possible outcome of the first event. If branches for a second event are to be drawn, the results of the second event must be drawn from *every* outcome of the first event. With multiple events, tree diagrams can get very big very quickly, but they can be helpful in determining the sample space or probability of an experiment.

EXAMPLE 4: Let's look at a very simple tree diagram—one that describes the events that can happen when flipping a single coin:

EXAMPLE 5: Typically, we will use tree diagrams to illustrate situations in which more than one event occurs. For example, flipping a coin two times:

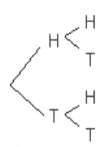

This endpoint represents: first flip heads, second flip heads.

This endpoint represents: first flip heads, second flip tails.

This endpoint represents: first flip tails, second flip heads.

This endpoint represents: first flip tails, second flip tails.

The above tree diagrams are read left to right, one event at a time. What we will call a "path" is any route that you can take to get from the beginning point on the tree to an endpoint of the tree, and going backwards is not allowed. The diagram that describes a single coin flip has two different paths, one that goes to H, and one that goes to T. The diagram that describes two different coin flips is more complicated, and has four different paths:

1. From the beginning point, go to Heads and then to Heads—you have reached an endpoint and are at the end of a path.

2. From the beginning point, go to Heads and then to Tails—you have reached an endpoint and are at the end of a path.

3. From the beginning point, go to Tails and then to Heads—you have reached an endpoint and are at the end of a path.

4. From the beginning point, go to Tails and then to Tails—you have reached an endpoint and are at the end of a path.

Yes, the number of endpoints that you have will be the same as the number of paths, and this is the number of elements that are in the sample space. For this experiment, the sample space that comes directly from the paths we just described is {HH, HT, TH, TT}.

Trees & Probabilities

Finding a probability using a tree diagram, at this point, is a matter of putting the number of successes over the number of outcomes.

EXAMPLE 6: When flipping two coins, what is the probability that both coins are tails?

The number of outcomes shown in the tree (and in our sample space) is 4. That will be the denominator of our answer.

The number of successes (find the endpoints that have both flips as tails) is 1. That will be the numerator of our answer.

So, the probability of getting two tails is P(TT) = 1/4.

EXAMPLE 7: When flipping two coins, what is the probability that at least one of the coins is heads?

The number of outcomes shown in the tree is 4. That will be the denominator of our answer.

The number of successes (find the endpoints that have at least one occurrence of heads) is 3. That will be the numerator of our answer.

So, the probability of getting at least one head is P(at least one H) = 3/4

As the problems get more complicated, the probabilities of the events are written directly on the tree diagram. Since the tree is a picture of successive events, the probabilities are multiplied together as we move along a path.

EXAMPLE 8: If we flip a coin two times, what is the probability of getting two heads?

To solve this problem, we will draw a tree diagram and include the probability of each event.

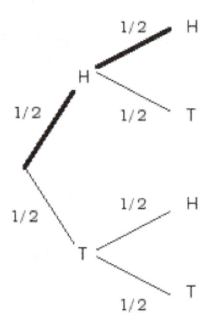

To arrive at the outcome of two heads, we must get heads on the first flip (in bold) *and* heads on the second flip (also in bold). Since these are successive events in our tree, we will multiply their probabilities together to find the probability of getting two heads. P (2 heads) = 1/2 × 1/2 = 1/4

Sometimes the problems will be even more complicated, and a tree diagram will become extremely useful. In the following example, it is the combination of the "at least" condition and the fact that experiment is done *without replacement* that causes the need for the tree. Let's go ahead and take a look at a problem that is fairly complicated...

EXAMPLE 9: We are going to buy two gumballs from a machine that contains five white and two red gumballs. What is the probability that we get *at least* one red gumball?

First, we draw the tree diagram for the situation. The first gumball will either be red or white. The second gumball will either be red or white.

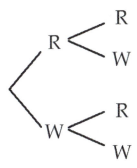

Next, we put the probabilities for the first event on the diagram. Since there are seven total gumballs, five white, and two red, the probabilities go on as follows:

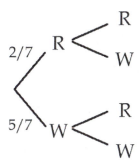

Since buying gumballs is done *without replacement*, we have to think carefully before placing the next probabilities on the diagram. If the first ball is red (following the 2/7 path), there are six gumballs left, and five of them are white, but only one of them is red (we bought one of the red ones already). If the first ball is white (following the 5/7 path), there are six gumballs left, but now two red ones remain and only four of the whites (we bought a white one). Now we insert those probabilities:

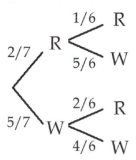

Since these are successive events, we multiply the probabilities to arrive at each outcome:

$$\begin{array}{l} 2/7 \nearrow R \begin{cases} 1/6 \; R \\ 5/6 \; W \end{cases} \\ 5/7 \searrow W \begin{cases} 2/6 \; R \\ 4/6 \; W \end{cases} \end{array}$$

2/42 is the probability of red, red.

10/42 is the probability of red, white.

10/42 is the probability of white, red.

20/42 is the probability of white, white.

Now we are ready to answer the question. We were asked for the probability of getting *at least* one red, so we have to consider every one of the outcomes that satisfies this condition.

$$\begin{array}{l} 2/7 \nearrow R \begin{cases} 1/6 \; R \\ 5/6 \; W \end{cases} \\ 5/7 \searrow W \begin{cases} 2/6 \; R \\ 4/6 \; W \end{cases} \end{array}$$

2/42 is the probability of red, red.

10/42 is the probability of red, white.

10/42 is the probability of white, red.

20/42 is the probability of white, white.

We have three different outcomes that satisfy the condition of the problem (getting at least one red gumball), which are the outcomes RR, RW, and WR. Since each of these outcomes represents a way to have a success, we add the probabilities of these outcomes together to find that the probability of getting at least one red gumball is:

(2/42) + (10/42) + (10/42) = 22/42

This means a little over half the time, we would get at least one red gumball.

That gumball question is not easy. If you are comfortable with that one, you should be just fine with all of the tree diagram problems in this course.

Finally, we will look at a tree diagram that is read top to bottom. Either method of drawing them is acceptable.

EXAMPLE 10: Create a tree diagram, which will allow you to determine the sample space (list of all possible outcomes) for a pair of six-sided dice that are rolled.

This is what your diagram should look like: (at the end, it shows the same sample space as your textbook for rolling a pair of dice)

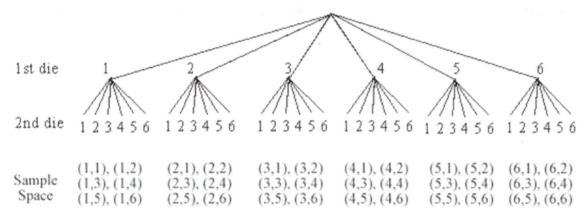

NOTE: (1,3) is the outcome of rolling a 1 on the first die and a 3 on the second die.

What are Tree Diagrams Good For?

Tree diagrams are a nice way to determine a sample space and, thus, the theoretical probability of each outcome. Use the tree diagram we just created to answer the following.

If two dice are rolled, how many outcomes are possible?

Assuming an event is the sum of the top faces of the dice:

- How many events create a total of seven? What is the probability of the sum being seven?

- How many events create a total of five? What is the probability of the sum being five?

- How many events create a total of three? What is the probability of the sum being three?

- How many events create a total of two? What is the probability of the sum being two?

- How many events create a total of one? What is the probability of the sum being one?

Section 4.4 Exercises

1. A jar of jellybeans contains seven red, two orange and three green jellybeans. One is selected and replaced, and then, another is selected. Find the probability that:

   a. The first jellybean is orange and the second is green.

   b. The first jellybean is red and the second is orange.

   c. The first jellybean is green and the second is red.

   d. Both jellybeans are orange.

   e. Both jellybeans are red.

   f. Both jellybeans are green.

2. A card is selected from a standard 52-card deck, and then, another card is selected without replacement. Find the probability that:

   a. The first card is a jack, and the second is an ace.

   b. The first card is the jack of hearts, and the second is an ace.

   c. The first card is a queen, and the second is also a queen.

   d. The first card is a spade, and the second card is also a spade.

3. The teachers that will have bus duty are drawn randomly each week. Mr. Art, Mrs. Biology, Ms. Chemistry, Mr. Drama, and Mrs. English have their names put into a hat. Three names will be drawn out without replacement. Find the probability that:

    a. Ms. Chemistry is chosen first.

    b. Mrs. Biology or Mr. Drama is chosen first.

    c. Ms. Chemistry is chosen first, Mr. Art is chosen second, and Mrs. English is chosen third.

    d. Mr. Drama is chosen first and second.

4. A bag contains 11 blue, 7 white, 4 red, and 6 purple marbles.

    a. Find the probability that a single draw results in a white marble.

    b. Find the probability of getting two blue marbles on two draws, with replacement.

    $$\frac{11}{28} \times \frac{11}{28}$$

    c. Find the probability of getting two blue marbles on two draws, if replacement is not allowed.

    d. Find the probability of getting two red marbles on two draws, if replacement is not allowed.

e. Find the probability of getting a white and, then, a purple marble, if replacement is not allowed.

5. A family will decide where they should have dinner randomly and have the following options:

Location:	City, Suburbs, Rural
Style:	Sit-down, Buffet
Price:	Expensive, Moderate, Cheap

a. What is the probability that they will eat in the city?

b. What is the probability that they will eat at a buffet?

c. What is the probability that they will eat at a cheap restaurant?

d. What is the probability that will eat at an expensive, sit-down restaurant in the suburbs?

e. What is the probability that they will eat at a rural, sit-down restaurant that is either expensive or moderate.

6. A nickel is flipped and then it is flipped again.

   a. Use a tree diagram to show the sample space and the probabilities.

   b. Find the probability that both flips result in heads.

   c. Find the probability that both flips result in tails.

   d. Find the probability that the first flip is heads and the second flip is tails.

   e. Find the probability that at least one heads is obtained.

   f. Find the probability that neither coin results in heads.

7. A family consists of four boys and two girls. Two of them will be picked, at random, to do yard work this weekend.

   a. Use a tree diagram to show the sample space and the probabilities.

   b. Find the probability that both people that will be doing yard work are girls.

   c. Find the probability that none of the people doing yard work are girls.

d. Find the probability that exactly one of the people doing yard work is a girl.

8. A gumball machine contains six red and four blue gumballs. Two of them are purchased (without replacement).

a. Use a tree diagram to show the sample space and the probabilities.

b. Find the probability that at least one of the gumballs is red.

c. Find the probability that exactly one of the gumballs is red.

d. Find the probability that at least one of the gumballs is blue.

e. Find the probability that exactly one of the gumballs is blue.

f. Find the probability that the second gumball is red.

9. What is the difference between with and without replacement?

Answers to Section 4.4 Exercises

Fractional answers that are reduced are acceptable, but the fractional answers given here have not been reduced.

1. a. 6/144   b. 14/144   c. 21/144   d. 4/144
    e. 49/144   f. 9/144

2. a. 16/2652   b. 4/2652   c. 12/2652   d. 156/2652

3. a. 1/5   b. 2/5   c. 1/60   d. 0

4. a. 7/28   b. 121/784   c. 110/756   d. 12/756
    e. 42/756

5. a. 6/18   b. 9/18   c. 6/18   d. 1/18
    e. 2/18

6. a. H (1/2) – H (1/2) = HH (1/4)
        H (1/2) – T (1/2) = HT (1/4)
        T (1/2) – H (1/2) = TH (1/4)
        T (1/2) – T (1/2) = TT (1/4)
    b. 1/4   c. 1/4   d. 1/4
    e. 3/4   f. 1/4

7. a. B (4/6) – B (3/5) = BB (12/30)
        B (4/6) – G (2/5) = BG (8/30)
        G (2/6) – B (4/5) = GB (8/30)
        G (2/6) – G (1/5) = GG (2/30)
    b. 2/30   c. 12/30   d. 16/30

8. a. R (6/10) – R (5/9) = RR (30/90)
        R (6/10) – B (4/9) = RB (24/90)
        B (4/10) – R (6/9) = BR (24/90)
        B (4/10) – B (3/9) = BB (12/90)
    b. 78/90   c. 48/90   d. 60/90
    e. 48/90   f. 54/90

9. When we sample "with replacement," the two sample values are independent. This means the result of the first event does not affect the second event. For "without replacement," the two sample values are not independent. This means what we get for the first event does affect what we can get for the second one.

4.5 What It's Worth (Expected Value)

In the Long Run...

Remember that phrase: In the long run ... Although probabilities are used to make a prediction as to what should happen for one particular event, the expected value calculations are all based on the long-term expectations of an event. While we expect one out of every two flips of a coin to be heads, sometimes we might get five tails in a row, so there are no guarantees in the short term. What a probability value tells us is that while a few flips might give strange results, if we flip that coin 1,000 times, it is very likely that the number of heads we end up with will be fairly close to 500. This idea is known as **The Law of Large Numbers**.

Expected Value

The **expected value** calculation is used to determine the value of a business venture or a game over the long run.

> **Expected value** is *not* the value of a particular event.
> It is the long run average value if the event was repeated many times.

Assume there is a game in which we either win $40 or lose $1, and we play the game 20 times, winning twice. How much money can we expect to have won or lost at the end of those 20 turns?

$$2 \text{ wins} \times \$40 + 18 \text{ losses} \times (-\$1) = \$80 + (-\$18) = \$62$$

Since we played 20 times, the average amount we won per turn was $62/20 = $3.10. In other words, the long run results would have been the same as if we had won $3.10 each time we played. $3.10 is the expected value of each turn for that game.

We have a formula to calculate the expected value (EV) of these types of events, which uses the possible results and the probability of each of those results:

$$EV = (\text{prob \#1}) \times (\text{result \#1}) + (\text{prob \#2}) \times (\text{result \#2}) + ... + (\text{last prob}) \times (\text{last result})$$

The game discussed above had two results, win $40 or lose $1. To find the expected value, we will need the following information:

Result #1 = win $40	Probability of result #1 = 2/20
Result #2 = lose $1	Probability of result #2 = 18/20

Plugging these values into our formula, we have:

$$EV = (\text{prob \#1}) \times (\text{result \#1}) + (\text{prob \#2}) \times (\text{result \#2})$$

$$EV = (2/20) \times (\$40) + (18/20) \times (-\$1) = \$4 - \$0.90 = \$3.10$$

Let's find the expected value of that same game, with one small change. Let's say, this time, when we win, we win $10. Also, this time, just to show it can be done either way, we will use decimal values for the probabilities:

$$EV = (\text{prob \#1}) \times (\text{result \#1}) + (\text{prob \#2}) \times (\text{result \#2})$$

$$EV = (0.1) \times (\$10) + (0.9) \times (-\$1) = \$1 - \$0.90 = \$0.10 = 10¢$$

This result is quite a bit different from the other set up. This time, over the long run, we will only win 10 cents per play.

Let's find the expected value of this same game one more time. This time, when we win, we only win $5:

$$EV = (\text{prob \#1}) \times (\text{result \#1}) + (\text{prob \#2}) \times (\text{result \#2})$$

$$EV = (0.1) \times (\$5) + (0.9) \times (-\$1) = \$0.50 - \$0.90 = -\$0.40 = -40¢$$

Notice the version with a $5 prize has a *negative* expected value. That means we would expect to *lose* an average of 40 cents per play. If we were to play this game 20 times, we would expect to lose a total of $8, (20 × −$0.40).

Expected Value, Revisited

Consider the following game, in which a spinner is used for play. Calculate the expected value of the spinner:

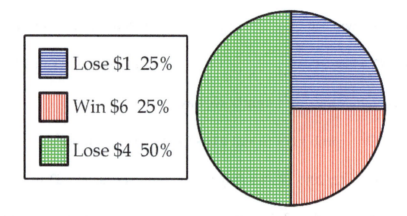

In this game, we have three possible results; lose $1, win $6, or lose $4. To find the expected value, we will use the following:

Result #1 = lose $1   Probability of result #1 = 1/4

Result #2 = win $6   Probability of result #2 = 1/4

Result #3 = lose $4   Probability of result #3 = 1/2

Using the formula for expected value,

EV = (prob #1) × (result #1) + (prob #2) × (result #2) + ... + (last prob) × (last result)

EV = (1/4)(−1) + (1/4)(+6) + (1/2)(−4)

EV = (–1/4) + (6/4) + (–4/2)

EV = (–1/4) + (6/4) + (–8/4)

EV = –3/4 = –$0.75

The negative expected value tells us that we will lose money in this game

Keep in mind that this does *not* mean we will play one time and lose 3/4 of a dollar, or 75 cents. What this expected value means is, over the long term, we will lose $3 for every four times we play, for an average loss of 75 cents per play.

EXAMPLE 1: Find the expected value of the spinner.

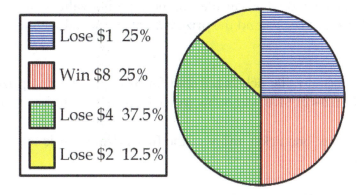

We begin by finding each of our results, and the probabilities of each of those results. While we could find this expected value by using decimals, here, we will use fractions:

Result #1 = lose $1   Probability of result #1 = 1/4

Result #2 = win $8   Probability of result #2 = 1/4

Result #3 = lose $4   Probability of result #3 = 3/8

Result #4 = lose $2   Probability of result #4 = 1/8

From there, we can use the formula for expected value:

EV = (1/4)(–1) + (1/4)(+8) + (3/8)(–4) + (1/8)(–2)

EV = (–1/4) + (8/4) + (–12/8) + (–2/8)

EV = (–1/4) + (8/4) + (–6/4) + (–1/4)

EV = 0/4 = $0

Raffles are another kind of game, and in a raffle, the player must pay up-front for the opportunity to take part. The money paid for a raffle ticket is not added on to the prize money, and that has to be considered when calculating the expected value of a raffle. For example, if a player pays $5 for a raffle ticket and the prize is $100, the actual winnings would be $95, not $100.

EXAMPLE 2: One thousand raffle tickets are sold for $1 each. One first prize of $500 and two second prizes of $100 will be awarded. Consider the following questions.

   a. If we buy one ticket and win the first prize, what is our *net* profit? (It is not $500!)

   Answer: $499

   b. What is the probability of winning the first prize?

   Answer: 1/1000 or 0.001

   c. If we buy one ticket and win one of the second prizes, what is our *net* profit?

   Answer: $99

   d. What is the probability of winning $100?

   Answer: 2/1,000 or 0.002

   e. If we buy one ticket and win nothing, what is our *net* profit?

   Answer: –$1 (You lose $1)

f. What is the probability of winning nothing at all?

Answer: 997/1,000 or 0.997

g. Use the above information to find the expected value of each $1 ticket?

Answer: EV = 0.001($499) + 0.002($99) + 0.997(–$1) = –$0.30 or –30¢

h. What are the possible financial outcomes for us if we purchase one ticket?

Answer: Win $499, Win $99, or Lose $1

i. If we were to buy all of the tickets, what would our expectation be (in a dollar amount)?

Answer: We would win $700, but spend $1000. The expectation is -$300.

j. What are the possible financial outcomes for us if we purchase all of the tickets?

Answer: We will lose $300.

k. If we were to buy 5 tickets, what would the expectation be (in a dollar amount)?

Answer: –$0.30(5) = –$1.50

l. What are the possible financial outcomes for us if we purchase five tickets?

Answer: Win $695 (first prize and both second prizes, minus the $5 ticket cost.)
Win $595 (first prize and one second prize, minus the $5 ticket cost.)
Win $495 (first prize, minus the $5 ticket cost.)
Win $195 (Both second prizes, minus the $5 ticket cost.)
Win $95 (One second prize, minus the $5 ticket cost.)
Lose $5

Fair Games

A game is considered "fair"—in the mathematical sense—when the expected value is equal to zero, meaning we should break even over the long run, and there is no advantage for either player. The spinner from Example 1 is an example of a fair game.

Fair Price

The **fair price** to pay for playing a game can be found by calculating the expected winnings. Our formula here is nearly identical to the one used to find expected value, as we are looking for the amount we would expect to win from playing this game.

Exp Winnings = (prob #1) × (result #1) + (prob #2) × (result #2) + ... + (last prob) × (last result)

EXAMPLE 3: A high school is selling 500 raffle tickets for a free dinner for two, valued at $75. What would be a fair price to pay for one ticket?

Expected Winnings = (1/500)x($75) = $0.15, so the fair price to pay for a ticket would be 15 cents. Naturally, the school would not sell them for 15 cents each, but likely for $1 each, so that they can make a profit on the raffle.

EXAMPLE 4: A carnival game allows a player to draw a card out of a hat. There are three cards inside the hat that correspond to the prizes listed below.

• Ace of spades = Win $10

• King of diamonds = Win $5

• Queen of hearts = Win $1

What is a fair price to pay for playing this game?

As we did when calculating expected values, we must determine the possible results and the corresponding probabilities. Then, we can determine the amount we would expect to win when playing this game.

Result #1 = win $10   Probability of result #1 = 1/3

Result #2 = win $5   Probability of result #2 = 1/3

Result #3 = win $1   Probability of result #3 = 1/3

Expected Winnings = (1/3) × ($10) + (1/3) × ($5) + (1/3) × ($1)

Expected Winnings = ($10/3) + ($5/3) + ($1/3) = $16/3 = $5.33

Since the expected winnings is $5.33, the fair price to pay to play this game would be $5.33. Of course, if the carnival wants to make a profit, they will charge more than this.

Keeping in mind a fair price for a game is the amount that makes the expected value zero. If the cost to play is less than the expected value, then the person playing the game can expect to make a profit. Thus, another way to compute the expected value is to compute the fair price and, then, subtract the cost to play at the end.

Life Insurance

Buying life insurance is an important step in the lives of many people. If the policyholder passes away during the term of the insurance policy, the amount of the policy is paid to the beneficiaries.

In order to establish the cost of a life insurance policy, insurance companies do research to determine the likelihood of death at different ages for both males and females. A table of these probabilities is created to determine the rates and can be used to calculate the expected value of purchasing a policy. Finding the expected value of a life insurance policy will share one similarity with raffles. Since the cost of the policy is not returned to the policyholder, that cost must be subtracted from the payout.

EXAMPLE 5: A 47-year old woman has a 0.248% chance of passing away during the next year. An insurance company charges $300 for a life insurance policy that pays a $110,000 death benefit. What is the expected value for the person buying the insurance?

To find this expected value, we will list the possible results and corresponding probabilities.

Keep in mind, if the probability of the policyholder passing away is 0.248% (which is 0.00248), then the probability of the policyholder *not* passing away is $1 - 0.00248 = 0.99752$.

Result #1 = Policyholder does not pass away (−300)   Prob of result #1 = 0.99752

Result #2 = Policyholder does pass away (109,700)   Prob of result #1 = 0.00248

Using the formula for expected value,

$$EV = (\text{prob } \#1) \times (\text{result } \#1) + (\text{prob } \#2) \times (\text{result } \#2)$$

$$EV = (0.99752)(-300) + (0.00248)(109{,}700)$$

$$EV = -299.256 + 272.056$$

$$EV = -27.20$$

As would be the case with auto insurance, it isn't really a surprise that the expected value of a life insurance policy is negative. After all, the insurance company can only afford to stay open (and offer insurance) if it is making a profit. When it is in the position to pay out a benefit, the large number of policies that it carries offsets the cost of this payment.

For those who purchase life insurance, even though there is a negative expected value, this cost is acceptable given the security that comes along with having the life insurance policy in place.

Section 4.5 Exercises

1. Given the spinner on the right:

    a. Find the expected value.

    b. What does this expected value mean?

2. Given the spinner on the right:

    a. Find the expected value.

    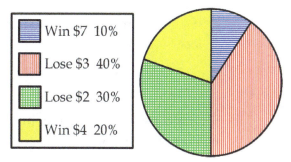

    b. What does this expected value mean?

3. A game consists of a player drawing a single card from a standard deck. If the card is a spade, the player wins $5; if the card is a club, the player wins $2; and if the card is red, the player loses $4.

    a. What is the expected value for someone who plays this game?

    b. What does this expected value mean?

    c. If someone played this game 100 times, how much money should he or she expect to win or lose?

4. An American roulette wheel contains slots with numbers from 1 through 36 and slots marked 0 and 00. Eighteen numbers are colored red, and eighteen numbers are colored black. The 0 and 00 are colored green. You place a $1 bet on a roulette wheel betting that the result of the spin will be green. If you win, the amount of your winnings is $17.

   a. What is the expected value of this bet? Round your answer to the nearest hundredth.

   b. What does this expected value mean?

   c. If someone made this bet 100 times, how much money should he or she expect to win or lose?

5. Scott is considering investing in a stock that he heard about. There is a 35% chance that he will lose $6,000, a 45% chance that he will break even (make $0), and a 20% chance that he will make $11,000.

   a. What is the expected value of this investment?

   b. Should Scott invest in this stock?

6. An urn contains 7 red, 10 white, and 3 blue marbles. A game involves drawing one marble from the urn, and if the marble is red, the player wins $20. If the marble is white, the player loses $10, and if the marble is blue, the player wins $40. What is a fair price to pay for playing this game?

7. Four baseball caps are on a table. One cap has a $1 bill under it, another has a $5 bill, another has a $50 bill, and the last one has a $100 bill. You can select one cap, and keep the amount of money that is underneath. What is the fair price for playing this game?

8. A game involves a player drawing a single card from a standard deck. If the card has an even number on it, the player wins $4. If the card is a face card (jack, queen or king), the player wins $5. If any other card is drawn, the player loses. In order for this to be a fair game, when a player loses, how much should he lose?

9. Five hundred tickets will be sold for a raffle, at a cost of $2 each. The prize for the winner is $750. What is the expected value of a ticket for someone playing this raffle?

10. Eight hundred raffle tickets will be sold for $5 each. One first prize of $500 will be awarded, along with one second prize of $100. What is the expected value for a person who buys a single ticket for this raffle?

11. According to the tables used by insurance companies, a 36-year old man has a 0.166% chance of passing away during the coming year. An insurance company charges $260 for a life insurance policy that pays a $150,000 death benefit. What is the expected value for the person buying the insurance?

Answers to Section 4.5 Exercises

1. a. − $1.15
    b. That, over the long term, you can expect to lose $1.15 for each spin.

2. a. − $0.30
    b. That, over the long term, you can expect to lose 30 cents for each spin.

3. a. −$0.25   b. Lose 25¢ per play   c. Lose $25

4. a. −$0.05
    b. That, over the long term, you can expect to lose 5 cents for each spin.
    c. Lose $5

5. a. Make $100   b. Yes, based upon expected value.

6. $8

7. $39

8. $7

9. −$0.50

10. −$4.25

11. −$11

Credits

1. "Dice," http://pixabay.com/en/dice-game-luck-gambling-cubes-red-161377/. Copyright in the Public Domain.

2. Copyright © Wikimedia Foundation, Inc. (CC BY-SA 3.0) at http://en.wikipedia.org/wiki/Blaise_Pascal.

3. Janmad, "Blaise Pascal," http://commons.wikimedia.org/wiki/File:Blaise_Pascal_Versailles.JPG. Copyright in the Public Domain.

4. Copyright © Wikimedia Foundation, Inc. (CC BY-SA 3.0) at http://en.wikipedia.org/wiki/Pierre_de_Fermat.

5. François de Poilly, "Portrait of Pierre de Fermat," http://commons.wikimedia.org/wiki/File:Pierre_de_Fermat_(F._Poilly).jpg. Copyright in the Public Domain.

6. Copyright © Wikimedia Foundation, Inc. (CC BY-SA 3.0) at http://en.wikipedia.org/wiki/Gerolamo_Cardano.

7. Copyright © Wellcome Images (CC by 4.0) at http://wellcomeimages.org/indexplus/image/V0001004.html.

8. Copyright © Zoetnet (CC by 2.0) at http://commons.wikimedia.org/wiki/File:Caf%C3%A9_Marly,_Paris_2010.jpg.

9. Copyright © Dustin M. Ramsey (CC BY-SA 2.5) at http://commons.wikimedia.org/wiki/File:License_plate_shed_near_Parrsboro,_NS_-_08659.jpg.

10. "Dice," http://pixabay.com/en/dice-die-cube-gambling-games-3d-149215/. Copyright in the Public Domain.

11. Patpitchaya, "Heads or Tails," http://www.freedigitalphotos.net/images/Casino_and_gambling_g213-Heads_Or_Tails_p70720.html. Copyright © by FreeDigitalPhotos.net. Reprinted with permission.

12. "Dice," http://pixabay.com/en/dice-die-cube-gambling-games-3d-149215/. Copyright in the Public Domain.

13. "Ace of Spades," http://pixabay.com/en/spades-ace-card-playing-deck-297839/. Copyright in the Public Domain.

14. Copyright © Baishampayan Ghos (CC by 2.0) at http://commons.wikimedia.org/wiki/File:Las_Vegas_sportsbook.jpg.

15. "Plastic Balls," http://pixabay.com/en/plastic-balls-balls-colorful-color-456608/. Copyright in the Public Domain.

16. "Jelly Beans," http://pixabay.com/en/jelly-beans-candy-sugar-sweet-3150/. Copyright in the Public Domain.

17. "Marble," http://pixabay.com/en/marble-balls-colorful-single-one-1943/. Copyright in the Public Domain.

18. "Clemens XI Coin," http://pixabay.com/en/coin-embossing-pope-clement-xi-62935/. Copyright in the Public Domain.

19. Stuart Miles, "Admission Tickets," http://www.freedigitalphotos.net/images/Other_leisure_activi_g229-Admission_Tickets_p53882.html. Copyright © by FreeDigitalPhotos.net. Reprinted with permission.

20. "Baseball Cap," http://pixabay.com/en/cap-baseball-hat-isolated-visor-304059/. Copyright in the Public Domain.

5

GEOMETRY

LET NO MAN IGNORANT OF GEOMETRY ENTER HERE

That phrase was inscribed over the door to **Plato's Academy** in Athens. The ancient Greeks were the pioneers of geometry. Before them, the Egyptians and the Babylonians used many geometric principles in practical applications, but the Greeks studied its philosophical properties.

Traditionally, in most high school geometry classes, students are subjected to numerous two-column proofs, which are not the most fun or interesting things in the world. While there is merit in learning those proofs and techniques, the beauty and extreme usefulness of geometry gets lost. Whether we are looking at the circular wheels on a bicycle or the straight lines and right angles of walls, ceilings, and doorways, our world would be a very different place without a basic understanding of geometry.

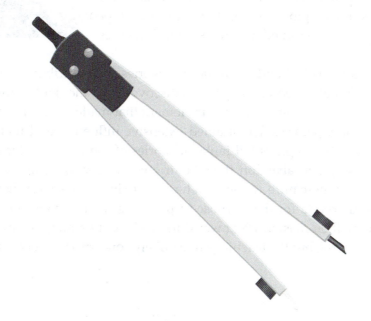

5.1 On the Shoulders of Giants (Biographies & Historical References)

The ancient Greeks were the pioneers of geometry. Even though many of the writings that exist today were originally collaborative efforts, a fair number of individuals stood out more prominently from the rest. Geometry, like all mathematics, progressed over time.

Pythagoras of Samos & the Pythagoreans

Pythagoras of Samos (569–475, BC) was a Greek philosopher, mathematician, and founder of the religious movement called Pythagoreanism. Most of the information about Pythagoras was written down centuries after he lived, so very little reliable information is known about him. He was born on the island of Samos, and might have travelled widely in his youth, visiting Egypt and other places seeking knowledge. Around 530 BC, he established a religious sect, and his followers pursued the religious rites and practices developed by Pythagoras and studied his philosophical theories.

Many mathematical and scientific discoveries were attributed to Pythagoras, including his famous theorem, as well as discoveries in the field of music, astronomy, and medicine. But it was the religious element that made the profoundest impression upon his contemporaries. He attained extensive influence, and many people began to follow him. Biographers tell fantastic stories of the effects of his eloquent speech in leading people to abandon their luxurious and corrupt way of life, and devote themselves to a more pure system that he came to introduce. His followers established a select brotherhood for the purpose of pursuing the religious and ascetic practices developed by their master. What was done and taught among the members was kept a profound secret, but the teachings most likely concerned science and mathematics.

248

The organization set up by Pythagoras was in some ways a school, in some ways a brotherhood, and in some ways a monastery. It was based upon the religious teachings of Pythagoras and was very secretive. The members were bound by a vow to Pythagoras and each other, for the purpose of pursuing the religious and ascetic observances, and of studying his religious and philosophical theories. Pythagoreans lived on a strict plant-based diet, and were even prohibited from eating beans. Considerable importance seems to have been attached to music and gymnastics in the daily exercises of the disciples. There were secret symbols, by which members of the sect could recognize each other, even if they had never met before. The society took an active role in politics, which eventually led to their downfall.

Exactly what was done and taught among the members was kept a profound secret. Candidates had to pass through a period of probation, in which their powers of maintaining silence were especially tested, as well as their general temper, disposition, and mental capacity. There were also gradations among the members themselves; it was an old Pythagorean maxim that every thing was not to be told to every body. Thus, the Pythagoreans were divided into an inner circle called the *mathematikoi* ("learners") and an outer circle called the *akousmatikoi* ("listeners").

Pythagoras made influential contributions to philosophy and religion in the late 6th century BC. He is often revered as a great mathematician, mystic, and scientist. However, because legend and secrecy cloud his work, some have questioned whether he personally contributed much to mathematics or natural philosophy. Since the achievements of any of the Pythagoreans are attributed to Pythagoras himself, it is possible that many of the accomplishments credited to Pythagoras may actually have been those of his colleagues and successors. Accurate facts about the life of Pythagoras are so few that it is nearly impossible to provide more than a vague outline of his life.[1]

Euclid's *Elements*

Very few original references to **Euclid of Alexandria** survive, so little is known about his life. The date, place and circumstances of both his birth and death are unknown, and may only be estimated relative to other figures mentioned alongside him. The few historical references to Euclid were written centuries after he lived, and he is rarely mentioned by other Greek mathematicians after the time of Archimedes.

Because this lack of biographical information is unusual for ancient Greek mathematicians, some researchers have proposed that Euclid was not, in fact, an actual historical figure, and that his works were written by a team of mathematicians who took the name Euclid from the Greek philosopher Euclid of Megara. That being said, this hypothesis is not well accepted by scholars and there is little evidence in its favor.

Elements is a series of 13 books attributed to Euclid, written around 300 BC. It is a collection of definitions, postulates, propositions, and mathematical proofs. The thirteen books cover the geometry that bears his name (Euclidean geometry), and the ancient Greek version of elementary number theory. The work also includes an algebraic system that has become known as geometric algebra, which is powerful enough to solve many algebraic problems, including the problem of finding the square root of a number (remember, this was written in 300 BC!). The *Elements* has proven instrumental in the development of logic and modern science, so much so that the theorems in it should be seen as standing in the same relation to geometry as letters are to language.

Euclid's *Elements* has been referred to as the most successful and influential textbook ever written. Being first set in type in 1482, it is one of the very earliest mathematical works to be printed after the invention of the printing press, and was estimated to be second only to the Bible in the number of editions published. For centuries, knowledge of at least part of Euclid's *Elements* was required of all students. Not until the 20th century, by which time its content was universally taught through other school textbooks, did it cease to be considered something to be read a by all educated people.[2,3]

Hypatia of Alexandria

The mathematician and philosopher **Hypatia of Alexandria** (370–415) was the daughter of the philosopher Theon. She was educated at Athens, and around AD 400, she became head of the Platonist school at Alexandria. Here, admired for her dignity and virtue, she imparted the knowledge of Plato and Aristotle to students, including pagans, Christians, and foreigners. Unfortunately, being an intelligent, dignified and forthright woman, just before the beginning of the Dark Ages, led some fanatical Christian sects to consider her teachings to be paganism.

Hypatia eventually became a focal point of conflict between Christians and non-Christians, which eventually led her to a violent death. Threatened by her knowledge, a Christian mob claimed that she beguiled many people through magic and satanic wiles, attacked her, dragged her from her carriage, tore off all her clothes and burned her to death. Many scholars believe her murder marked the beginning of the downfall of intellectual life in Alexandria.

No written work widely recognized by scholars as Hypatia's own has survived to the present time. Many of the works commonly attributed to her are believed to have been collaborative efforts between her and her father. This kind of authorial uncertainty is typical for female philosophers in antiquity.[4]

Indiana House Bill #246

In 1897, **Indiana House Bill No. 246** attempted to set the value of π to an incorrect rational approximation. The bill, written by amateur mathematician and medical doctor, Edwin Goodwin, was so poorly crafted; it even contradicted itself, implying three different values for π. Nevertheless, Dr. Goodwin succeeded in getting his State Representative, Taylor Record, to introduce the bill, under the agreement that the State of Indiana could use the information free of charge, but the rest of the Country would have to pay him royalties for its usage.

The bill, which, apparently, no Representative understood, passed through the Indiana House of Representatives unanimously (67–0), and was sent to the Senate for approval. Fortunately, during the House's debate on the Bill, Purdue University Mathematics Professor **Clarence Waldo** was present, and became horrified. After the debate, a Representative offered to introduce Professor Waldo to Dr. Goodwin. Prof. Waldo declined by stating that he was already acquainted with as many crazy people as he cared to know. Later that evening, Professor Waldo informed the members of the Indiana Senate of the "merits" of the bill. The next day, after some good-natured ridicule at the expense of their colleagues in the House, the Senate moved the bill to an obscure committee and let it die a painless death.

---

4   Copyright © Wikimedia Foundation, Inc. (CC BY-SA 3.0) at http://en.wikipedia.org/wiki/Hypatia.

5.2 How Tall is that Tree? (Right & Similar Triangles)

Square Roots

Given a quantity, the **principal square root** of that quantity is the positive number we must square to give that quantity. For our purposes, we will refer to the "principal square root" as just the "square root." For an example, 5 is a square root of 25, because

$5^2 = 25$. For notation, we write $\sqrt{25} = 5$.

For numbers like 25, 36, 81, and 100, finding the square roots is easily done by inspection. These numbers are **perfect squares**. How do you find the square root of a number that is not a perfect square? Although there are ways to do this by hand, the fastest way is to just use a calculator. Furthermore, unless directed otherwise, we will round square roots to the nearest hundredth.

The Pythagorean Theorem

The longest side of a right triangle is always the side opposite the right angle, and is called the **hypotenuse**. The other two sides are referred to as **legs**. The **Pythagorean Theorem** states:

In any right triangle, the square of the hypotenuse is equal to the sum of the squares of the legs.

That is, $\text{hypotenuse}^2 = \text{leg}^2 + \text{leg}^2$

EXAMPLE 1: Find the length of the hypotenuse in the following right triangle.

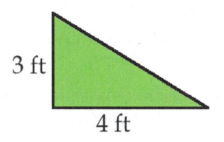

Since the legs measure 3 feet and 4 feet, the Pythagorean Theorem states:

$(\text{hypotenuse})^2 = 3^2 + 4^2$

$(\text{hypotenuse})^2 = 9 + 16$

$(\text{hypotenuse})^2 = 25$

Thus, the hypotenuse $= \sqrt{25} = 5$

EXAMPLE 2: Find the length of the missing side in a right triangle if the hypotenuse is 6 feet and one of the sides is 2 feet long.

We now know one of the sides and the hypotenuse, so the solving process will be a little different:

$6^2 = 2^2 + (\text{side})^2$

$36 = 4 + (\text{side})^2$

$32 = (\text{side})^2$

Thus, side $= \sqrt{32} = 5.6568... \approx 5.66\ ft$

What Good is the Pythagorean Theorem, Anyway?

Carpenters still use the 3-4-5 triangle to square corners. Here's how.

First, we need to think of a window frame and what happens to that frame when we anchor its bottom, but push the top to the left or the right.

In the figures on the right, a frame is shown in three states. The top frame is considered "square," because the angle X is a right angle.

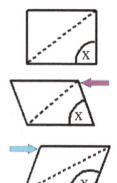

In the middle frame, the top has been pushed to the left. Notice the angle X is less than 90° and the dotted line is shorter than the same line in the top frame.

In the bottom frame, the top has been pushed to the right. Notice the angle X is greater than 90° and the dotted line is longer than the same line in the top frame.

Thus, when the bottom of the window frame is nailed down, we can adjust the length of the dotted line by pushing the top of the frame to the left or right

Then, we can be sure the window frame is "square" by a simple application of the Pythagorean Theorem. From a corner, we can make marks on the frame that are 3 inches in one direction and 4 inches in the other. Then, the whole frame can be adjusted until the straight-line distance between the two marks is 5 inches.

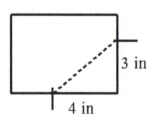

3 in

4 in

Essentially, since a triangle with a side ratio of 3-4-5 obeys the Pythagorean Theorem (that is, $3^2 + 4^2 = 5^2$), then that triangle *has* to be a right triangle. In other words, if you force one of these 3-4-5 triangles into a corner of a frame, then, according to the Pythagorean Theorem, the angle across from the hypotenuse (the side of length 5) has to be a right angle.

Right Triangle Trigonometry

Trigonometry. One of those things that seems pretty darn complicated. Trigonometry (also known as "trig") is used in several realms, including architecture, navigation, and engineering, but you know, if we break it down, it doesn't have to be mysterious. The word **trigonometry** is derived from the Greek words trigon (triangle) and metros (measure) and literally means "triangle measurement." Measuring a triangle, heck, that's not scary at all.

Although the science of trigonometry includes triangles of any type, we will limit our study to problems involving only right triangles, that is, triangles that contain a 90° angle.

Triangle Notation

Any time the side of a triangle is being referenced, a lowercase letter will be used. To denote an angle, an uppercase letter will be used. Furthermore, "a" will represent the side of the triangle that is across from angle "A," and so on.

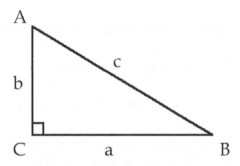

Opposite & Adjacent Sides of a Triangle

The terms **opposite** and **adjacent** play a very important role in right triangle trigonometry, so let's take a look at what those terms mean.

The three sides of a right triangle can be described in terms of each angle. For example, in the right triangle shown on the previous page, side "a" is across from, or **opposite** angle A. Side "b" is next to, or **adjacent** to angle A, and side "c" would be the hypotenuse.

Similarly, side "b" is **opposite** angle B, and side "a" is **adjacent** to angle B. Side "c" is still the hypotenuse.

The Trigonometric Ratios

There are three basic trigonometric functions, which are calculated by using the ratios of the lengths of the sides of the triangle. Those functions are:

Function	Pronunciation	Abbreviation
sine	sign	sin
cosine	co-sign	cos
tangent	tangent	tan

The sine of an angle is calculated using the following ratio:

sin A = (length of the side opposite angle A)/(length of the hypotenuse)

The cosine of an angle is calculated using the following ratio:

cos A = (length of the side adjacent to angle A)/(length of the hypotenuse)

The tangent of an angle is calculated using the following ratio:

tan A = (length of side opposite angle A)/(length of side adjacent to angle A)

This may seem like a lot to remember, but a really sharp teacher I had one time gave me a fancy word that I could use to help me remember it all. The "word" was: SOH CAH TOA.

SOH CAH TOA is shorthand for:

Sin = **O**pposite/**H**ypotenuse

Cos = **A**djacent/**H**ypotenuse

Tan = **O**pposite/**A**djacent

Let's put these functions into context by looking at an example:

EXAMPLE 3: For the triangle below, find the sine, cosine, and tangent for angles A and B. State your answers as fractions:

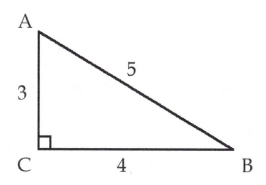

Answers:

sin A = 4/5	sin B = 3/5
cos A = 3/5	cos B = 4/5
tan A = 4/3	tan B = 3/4

Using Your Calculator

On any scientific calculator, you should see buttons labeled "sin," "cos," and "tan." These buttons will allow you to find the values of sine, cosine, and tangent for any angle measure Depending on the calculator, we may input the number and, then, push the "sin," "cos," or "tan" button, or we may have to push the trig button, type the number, and hit the "equals" button. **Also, make sure the calculator is in degree mode, not radian mode.** Let's try a few:

EXAMPLE 4: Use your calculator to find the following. Round your answers to four decimal places, if necessary:

a. sin 30°

d. cos 12°

b. sin 73°

e. tan 60°

c. cos 45°

f. tan 35°

Answers:

a. sin 30° = 0.5000 (or just 0.5)   d. cos 12° = 0.9781

b. sin 73° = 0.9563   e. tan 60° = 1.7321

c. cos 45° = 0.7071   f. tan 35° = 0.7002

NOTE: If you tried to find sin 30 and got -0.9880, your calculator is in radian mode.

Much like we did with the Pythagorean Theorem, we can use these trig ratios to find missing side lengths of a triangle. For instance, if we are given an angle measure and one of the side lengths, we can use that information to determine the length of the hypotenuse.

EXAMPLE 5: Given the following right triangle, find the measure of side "c." Round the value of any trig function to four decimal places, and then round your final answer to the nearest hundredth.

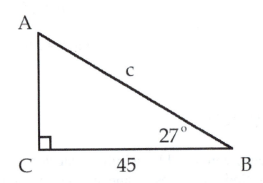

Solution:

Thinking back to SOH CAH TOA, the trigonometric function that uses adjacent and hypotenuse is cosine. So, we will use that function and, again, a little algebra, to solve this problem.

cos B = (length of side adjacent to angle B)/(length of the hypotenuse)

We can plug in the things we know, leaving only one variable, and, then, solve:

cos 27° = 45/c

Now, we can use our calculator to find cos 27° and round to four decimal places:

$$0.8910 = 45/c$$

Here, to get the "c" by itself will take two steps. First, we multiply both sides of the equation by "c." Since "c" is a distance and, therefore, not equal to zero, this step is legal.

$$0.8910(c) = 45$$

Finally, to get the "c" by itself, we divide both sides by 0.8910 to get c = 50.51

EXAMPLE 6: Use right triangle ABC, where side "c" is the hypotenuse, for the following problems. Round the value of any trig function to four decimal places, and then round your final answer to the nearest hundredth.

a. Given B = 14° and c = 20, find a.

b. Given A = 15° and a = 600, find c.

Solution:

To solve each of these, begin by drawing a triangle with side c as the hypotenuse (just like the triangle in Example 5).

In part (a), we are given angle B and side c, so label those in your diagram with the given numbers, and then, we need to find side a.

In relation to angle B, we are looking for the adjacent side and we know the length of the hypotenuse, so we should use cosine:

$$\cos 14 = a/20$$

Using our calculator, cos 14 = 0.9703, giving us:

$$0.9703 = a/20$$

Multiplying both sides by 20, we have:

$$19.41 = a$$

In part (b), we are given angle A and side a. Again, label those parts of your diagram, and then, we need to find side c.

In relation to angle A, we know the length of the opposite side and we are looking for the hypotenuse, so we should use sine:

$$\sin 15 = 600/c$$

Using our calculator, $\sin 15 = 0.2588$, giving us:

$$0.2588 = 600/c$$

Multiplying both sides by c, we have:

$$0.2588\, c = 600$$

Dividing both sides by 0.2588 gives us:

$$c = 2318.39$$

Let's put the usefulness of these trig functions into a more real-world context:

EXAMPLE 7: A 15-foot ladder is leaning against a wall so that it makes an angle of 62° with the wall. How far up the wall is the top of the ladder? Round the value of any trig function to four decimal places, and then round your final answer to the nearest hundredth of a foot.

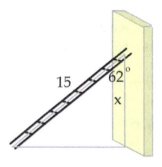

Solution:

Here, we are looking for the length of the side that is adjacent to the 62° angle, and we know the length of the ladder (which is the hypotenuse of the triangle).

The trig function that uses adjacent and hypotenuse is cosine, so we will use that function and, once again, a little algebra to find out how high up on the wall the ladder is resting.

$$\cos 62° = \text{(side adjacent to the 62° angle)}/\text{(length of the hypotenuse)}$$

We can plug in the length of the hypotenuse, and use a variable to represent the length of the side that we are trying to find. We can use any variable you like. This time we will use "x."

$$\cos 62° = x/15$$

Next, we can use our calculator to find cos 62° and round to four decimal places:

$$0.4695 = x/15$$

Finally, to get the "x" by itself, we multiply both sides by 15, giving us:

$$7.04 = x$$

So, the ladder is touching a spot on the wall that is 7.04 feet above the ground.

Angle of Elevation & Angle of Depression

An angle of elevation is an upward angle made between the line of sight and a horizontal line. Similarly, an angle of depression is a downward angle made between the line of sight and a horizontal line.

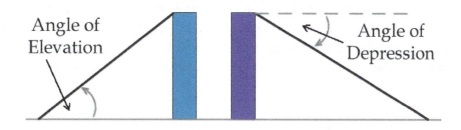

EXAMPLE 8: A 375-foot cable is attached to the top of a building, and the other end is anchored to the ground. The angle of elevation made by the cable, from ground level to the top of a building, is 29°. Find the height of the building. Round the value of any trig function to four decimal places, and then round your final answer to the nearest hundredth of a foot.

Solution:

Remember, the angle of elevation is the upward angle from the observer to the object:

In this problem, we are looking for the height of the building (the length of the side that is opposite the 29° angle), and we know the length of the cable (the length of the hypotenuse). The trig function that uses opposite and hypotenuse is sine, so we will use that function to solve this problem.

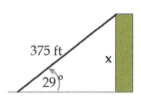

sin 29° = (side opposite the 29° angle)/(length of the hypotenuse)

We can plug in the length of the hypotenuse, leaving only one variable. Then, we can use our calculator to find sin 29°, and round to four decimal places. Finally, to get the "x" by itself, we multiply both sides by 375.

$$\sin 29° = x/375$$

$$0.4848 = x/375$$

$$181.80 = x$$

So, the building is 181.8 feet tall.

EXAMPLE 9: A spot on the ground below an airplane is 24,000 feet from the base of the control tower, and the angle of depression from the plane to the base of the control tower is 16°. Find the altitude of the plane. Round the value of any trig function to four decimal places, and then round your final answer to the nearest whole foot.

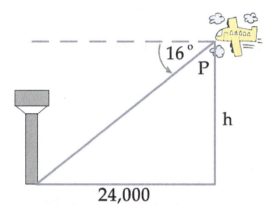

Solution:

In this problem, we are looking for the vertical distance from the plane to a spot on the ground below the plane (the length of the side that is adjacent to angle P), and we know the horizontal distance from a spot on the ground below the plane to the control tower (the length of the side opposite angle P).

We are not given the measure of angle P, but we can find it. We know the angle of depression is 16°, and the combination of angle P and the angle of depression makes a 90° angle. Therefore, we can determine that the measure of angle P is 90° − 16° = 74°.

The trig function that uses opposite and adjacent is tangent, so we will use that function to solve this problem.

tan P = (side opposite angle P)/(side adjacent to angle P)

tan 74° = 24000/h

Now, we can use our calculator to find tan 74 and round to 4 decimal places:

3.4874 = 24000/h

Here, to get the "h" by itself will take two steps. First, we multiply both sides of the equation by "h." Since "h" is a distance and, therefore, not equal to zero, this step is legal here.

3.4874(h) = 24000

$$h = 24000/3.4874 = 6881.92$$

So, the altitude of the plane is 6,882 feet.

Similar Triangles

Similar triangles are triangles whose angles have the same measure, but their sides have different lengths. If two triangles are similar, their side lengths are proportional to each other. The triangles will look identical, but one will be smaller than the other.

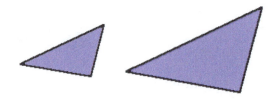

Using the proportionality of the sides, similar triangles are a simple and very powerful problem-solving tool. Using the correspondence of the angles of similar triangles, we can find the missing side lengths.

When setting up a proportion, each ratio in the proportion contains a pair of corresponding sides. Furthermore, we must be consistent with which triangle measures we place in the numerators and which ones we place in the denominators of the fractions. That is, think of the following layout when setting up a proportion.

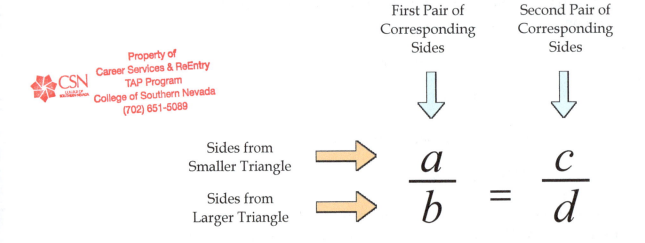

EXAMPLE 10: Given the following two triangles that are similar, we can see that angle A corresponds to angle E, angle B corresponds to angle F, and angle C corresponds to angle D. Find the lengths of BC and AC to the nearest hundredth.

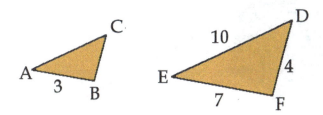

We can set up a proportion to find side BC: $\dfrac{AB}{EF} = \dfrac{BC}{FD}$

Plugging in the known numbers, we have: $\dfrac{3}{7} = \dfrac{BC}{4}$

Cross-multiplying gives us: 7(BC) = 12, so the side BC = 12/7 = 1.71

Similarly, we can set up a proportion to find side AC: $\dfrac{AB}{EF} = \dfrac{AC}{ED}$

Plugging in the known numbers, we have: $\dfrac{3}{7} = \dfrac{AC}{10}$

Cross-multiplying gives us: 7(AC) = 30, so the side AC = 30/7 = 4.29

Once again, similar triangles are extremely useful. We can actually use our own shadow to determine the height of a tree.

EXAMPLE 11: Let's say you are 5.5 feet tall, and, at a certain time of day, you find your shadow to be 10 feet long. At the same time of day, you measure the shadow of a tree to be 38 feet long. How tall is the tree? Here's a picture:

Draw the corresponding triangles on the figure, and label the sides we know.

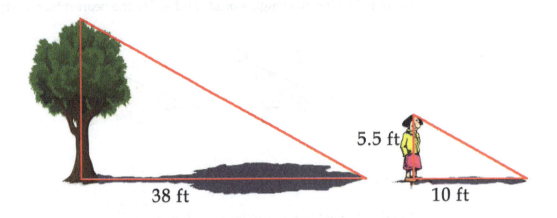

Then, set up a proportion that looks like:

$$\frac{Your\ Height}{Tree\ Height} = \frac{Your\ Shadow\ Length}{Tree\ Shadow\ Length}$$

$$\frac{5.5\ ft}{Tree\ Height} = \frac{10\ ft}{38\ ft}$$

Tree Height = (38 ft)(5.5 ft)/(10 ft) = 20.9 ft

So, the tree is 20.9 feet high. By the way, this is not 20 feet 9 inches—0.9 is 9/10 of a foot. 0.9 × 12 = 10.8, which is almost 11 inches.

The tree is almost 20 feet 11 inches tall.

Helpful Hint: To solve a problem involving similar triangles, a diagram is particularly helpful. If no diagram is provided in the question, draw one.

Section 5.2 Exercises

1. Find the following. Round to the nearest hundredth, as necessary.

   a. $\sqrt{36}$ b. $\sqrt{121}$ c. $\sqrt{63}$

2. If a right triangle has legs measuring 6 feet and 8 feet, how long is the hypotenuse?

3. If a right triangle has one leg that measures 12 centimeters, and a hypotenuse that measures 13 centimeters, how long is the other side?

4. If both legs of a right triangle measure 6 inches, how long is the hypotenuse? Round your answer to the nearest tenth of an inch.

5. If a 10-foot ladder is leaned against the top of a 9-foot high wall, how far will the base of the ladder be from the bottom of the wall? Round your answer to the nearest tenth of a foot.

6. If you drive 12 miles north, make a right turn, and drive 9 miles east, how far are you, in a straight line, from your starting point?

7. On a baseball diamond, there are 90 feet between home plate and first base and 90 feet between first and second base, and the base paths are at right angles. What is the straight-line distance from home plate to second base, to the nearest foot?

8. For a 25-inch television, the length of the screen's diagonal is 25 inches. If the screen's height is 15 inches, what is the width?

9. In the movie *The Wizard of Oz*, the Scarecrow proudly exclaims, upon being presented with his ThD (Doctor of Thinkology), "The sum of the square roots of any two sides of an isosceles triangle is equal to the square root of the remaining side." What he said was incorrect. Name at least three mistakes in the Scarecrow's statement.

10. For the triangle below, find the sine, cosine, and tangent for angles A and B. State your answers as fractions.

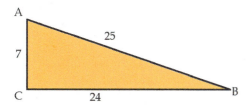

11. Use a calculator to find the following. Round your answers to four decimal places, if necessary.

    a. sin 38°   d. cos 33°

    b. sin 14°   e. tan 17°

    c. cos 71°   f. tan 63°

12. For the following, use right triangle ABC, where side "c" is the hypotenuse. Do not round the values of the trig functions, but round your final answer to the nearest hundredth.

    a. Given A = 53° and c = 97, find a.

    b. Given A = 41° and c = 200, find b.

    c. Given B = 85° and b = 110, find a.

    d. Given B = 72° and c = 300, find a.

    e. Given A = 28° and a = 872, find c.

f. Given A = 60° and b = 71, find a.

g. Given A = 30° and c = 40, find a.

h. Given B = 68° and a = 313, find c.

For Exercises #13 through #22, round all answers to the nearest hundredth. As you are solving the problem, round any values of trig functions to four decimal places.

13. A plane takes off from the ground at a 14° angle. After flying straight for 10,000 feet, what is the altitude of the plane?

14. A wire is tied to the top of a pole and attached to the ground at a spot 2 yards from the base of the pole. The wire makes a 64° angle with the ground. How long is the wire?

15. A security light is to be mounted on a building so that the angle of depression for the beam of light is 30°. How high must the light be mounted so that the light will shine on a spot that is 25 feet from the building?

16. A beetle is standing at a spot on the ground 22 feet from the base of a shed. If the angle of elevation from the beetle to the top of the shed is 27°, how tall is the shed? What is the straight-line distance from the beetle to the top of the shed?

17. Carl is standing on a platform built on the edge of a river that is 20 meters tall. The angle of depression from the top of the platform to the ground directly across the river is 53°. How wide is the river?

18. A 70-foot long cable is tied to the top of a building. Abigail is holding the cable to the ground so that the angle of elevation is 33°. How far is Abigail from the base of the building? How tall is the building?

19. The Stratosphere Tower is 350 meters tall. The angle of elevation from the spot where you are standing to the top of the Stratosphere Tower is 25°. How far from the base of the Stratosphere Tower are you?

20. The angle of depression from the top of a lighthouse to a ship is 17°. If the lighthouse is 140 feet tall, how far is the ship from the base of the lighthouse?

21. A ladder is leaning against a wall so that it makes an angle of 30° with the ground. The ladder touches the wall at a spot that is 8 feet from the ground. How long is the ladder?

22. A point on the edge of a canyon is 900 feet above the river below. The angle of depression to the middle of the canyon is 48°. What is the horizontal distance from the edge to a point directly above the middle of the canyon?

23. Given that the pictured triangles are similar, find the lengths sides DE and DF. Round your answers to the nearest tenth, as necessary.

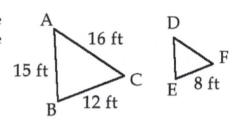

24. While standing 10 feet away from a light pole, Billy notices his shadow is 8 feet long. If Billy is 6 feet tall, how tall is the light pole?

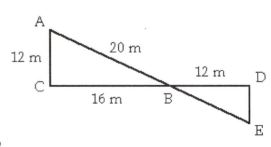

25. In the given figure, triangle ABC is similar to triangle EBD (denoted ΔABC ~ ΔEBD). What are the measures of BE and DE?

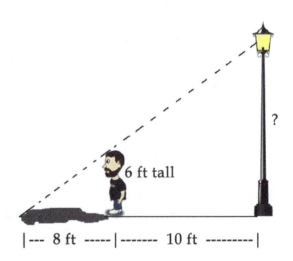

26. Mike wants to know the distance, w, across an east-to-west running river. He measures 120 feet along the southern bank of the river and then measures 16 feet south from the western edge of that measure (measure a in the figure). From that point, he cites a spot on the northern river bank that is straight north from the place where he started the 120-foot measure, and that line of sight crosses the original 120-foot line 32 feet from the western edge (measure b in the figure). This process creates two similar triangles as shown in the figure below. What is the distance across the river? Round your answer to the nearest foot

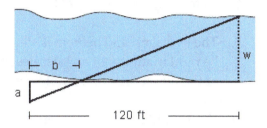

27. Two poles are 20 feet tall and 30 feet tall, respectively. If they are 30 feet apart, how far is it from the top of one pole to the top of the other pole? Round your answer to the nearest tenth of a foot.

273

Answers to Section 5.2 Exercises

1. a. 6   b. 11   c. 7.94

2. 10 ft   3. 5 cm   4. 8.5 in   5. 4.4 ft

6. 15 mi   7. 127 ft   8. 20 in

9. The Pythagorean Theorem is stated in squares, not square roots. The triangle must be right triangle, not an isosceles triangle. You have to take the sum of the squares of the two shorter sides, not "any" two sides.

10. $\sin A = 24/25$ $\sin B = 7/25$ $\cos A = 7/25$
    $\cos B = 24/25$ $\tan A = 24/7$ $\tan B = 7/24$

11. a. 0.6157 b. 0.2419   c. 0.3256   d. 0.8387   e. 0.3057   f. 1.9626

12. a. $a = 77.47$ b. $b = 150.94$ c. $a = 9.62$ d. $a = 92.71$
    e. $c = 1857.41$ f. $a = 122.98$ g. $a = 20.00$ h. $c = 835.54$

13. The altitude is 2,419 feet.   14. The wire is 4.56 yards long.

15. The light should be mounted 14.43 feet above the ground.

16. The shed is 11.21 feet tall, and the distance from the beetle to the top of the shed is 24.69 feet.

17. The distance across the river is 15.07 meters.

18. Abigail is 58.71 feet away from the building that is 38.12 feet tall.

19. You are 750.59 meters from the base of the tower.

20. The ship is 457.93 feet from the base of the lighthouse.

21. The ladder is 16 feet long.

22. The horizontal distance to a point above the center of the canyon is 810.36 feet.

23. DE = 10 ft, DF = 10.7 ft

24. Be careful. The base of the larger triangle is 18 feet, not just 10 feet. The lamppost is 13.5 feet high.

25. DE = 9 meters, BE = 15 meters

26. 44 ft   27. 31.6 ft

5.3 Getting into Shape (Polygons & Tilings)

A **closed broken line** is made up of line segments and begins and ends at the same point.

A **simple closed broken line** is one that does not intersect itself, and is known as a **polygon**.

Common (And Some Not-So-Common) Polygon Names

Name	Number of Sides	Picture
Triangle	3	
Quadrilateral	4	
Pentagon	5	
Hexagon	6	
Octagon	8	
Nonagon	9	
Decagon	10	

Name	Number of Sides	Picture
Dodecagon	12	
Triskaidecagon	13	
Icosagon	20	
Hectagon	100	

Notice how, as the number of sides of the polygon increases, the figure looks more and more like a circle. At the age of 19, **Carl Friedrich Gauss** proved it was possible to construct a regular **heptadecagon** (17 sides) using only a compass and a straightedge. He was so proud of the proof, he requested the shape be engraved onto his tombstone. The stonemason refused to perform the complicated task, stating the shape would have been indistinguishable from a circle.

The prefixes used in the polygon names define the number of sides. Tri- means three, quad- means four, penta- means five, and hexa-, octa-, and nona- mean six, eight, and nine, respectively. Deca- means 10, so a decagon has 10 sides. Can you recognize the prefixes for 12, 13, 20, and 100?

By the way, in the Roman Lunar Calendar, October was originally the 8th month, November (Novem- is an alternative prefix for nine) was originally the ninth month, and December was the tenth month. That original Roman calendar consisted of ten months, and began with March. **Julius Caesar** eventually reformed the calendar to 12 months and renamed a couple of mid-year months to honor himself and his nephew Augustus, whom he later adopted.

Types of Triangles & Their Properties:

Triangle	Properties	Picture
Equilateral	All sides are equal, all angles are equal	
Isosceles	At least two sides are equal	
Scalene	No sides are the same length, no angles are equal	
Right	One of the angles measures 90°	
Acute	All angles measure less than 90°	
Obtuse	One angle measures greater than 90°	

You may have noticed the same triangle appearing as an isosceles and an acute triangle. This does not mean all isosceles triangles are acute; it just means the pictured isosceles triangle is acute. We could easily have an isosceles triangle be obtuse. Likewise, the pictured scalene triangle is also obtuse, but that does not mean all scalene triangles are obtuse. Yes, a triangle may fall into more than one of the categories, but be sure treat each category according to its own properties.

Another subtle, but important concept appears in the picture of the right triangle. Notice that there is a small square appearing in the right angle of the triangle. Since all the angles in a square measure 90°, that small square is there to indicate that angle measures 90°.

The Triangle Inequality

In order to construct a triangle from three line segments, the sum of the measures of any two of the segments must be greater than the measure of the third segment. For example, we cannot construct a triangle with segments of lengths 3 centimeters, 4 centimeters, and 9 centimeters, because the sum of 3 and 4 is *not* greater than 9. If you don't believe it, grab a ruler, and try it!

Types of Quadrilaterals & Their Properties

Quadrilateral	Properties	Picture
Trapezoid	Exactly one pair of opposite sides is parallel.	
Parallelogram	Both pairs of opposite sides are parallel.	
Rectangle	All angles measure 90°.	
Rhombus	All sides are equal in length.	
Square	All sides are equal in length, and all angles measure 90°.	

Realize that many quadrilaterals may fall into more than one category. For example, a square satisfies the definition of a rhombus, rectangle, and parallelogram! To keep things simple, when we name a given quadrilateral, we should be as specific as possible. That is, even though all rectangles are parallelograms, if the angles all measure 90°, we should call it a rectangle—unless, of course, it is a square.

Are You Regular?

What does the word regular mean? In reference to a geometric figure, a **regular polygon** is a polygon whose sides are all the same length and whose interior angles are all the same measure. An **interior angle** an angle measured on the inside of a given polygon.

Which of the following figures are regular polygons? (Hint: Only two of them are regular.)

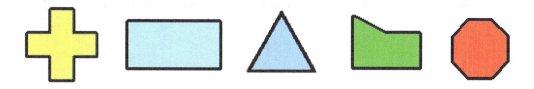

A well-remembered fact from any geometry course is that the sum of the interior angles for any triangle is 180°. This fact does not tell us the individual measures of the three angles; it only tells us their sum. Unless they are given or measured, the exact angle measures can only be determined if the triangle is regular. The interior angles of a regular triangle—also known as an equilateral triangle—are 60°, because all three angles are equal, and 180°/3 = 60°.

Any polygon—whether regular or not—can be broken into a finite number of triangles by cutting the polygon along the lines connecting non-consecutive angles. For example, a hexagon (six sides) can be thought of as a composition of four triangles.

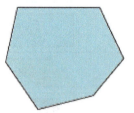 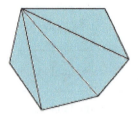 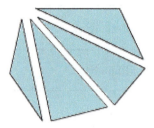

Since each triangle has an interior angle sum of 180° and a hexagon can be thought of as four triangles, the interior angle sum for any hexagon is 4(180°) = 720°. If the hexagon were a regular hexagon, then we could also say each of the six angles would be 120°.

The number of triangles composing any polygon is always two less than the number of sides of the polygon. Thus, the sum of the interior angles of any polygon can be determined by the formula (n−2)180°, where *n* is the number of sides for the polygon. Since, in a regular polygon, all the interior angles are equal, the total can then be divided by the number of angles to find the measure of each interior angle.

Keep in mind, this individual angle calculation can only be done with regular polygons.

EXAMPLE 1: What is the measure of each of the interior angles of a STOP sign?

First, we need to recognize a STOP sign is a regular octagon, which has eight sides. Then, the sum of the eight interior angles is (8−2)(180°) = (6)(180°) = 1,080°

Since all eight angles are of equal measure, each one measures 1,080°/8 = 135°

Tilings

If done correctly, a tiling can be a beautiful creation. A **tiling** also called a **tessellation**, is a space completely covered by geometric figures. A regular tiling is a tiling done exclusively with one particular regular polygon. Note the words "completely covered;" a tiling cannot contain spaces. Tilings can be done with a single figure or multiple figures, but in every case, there are no spaces. Here are a few examples.

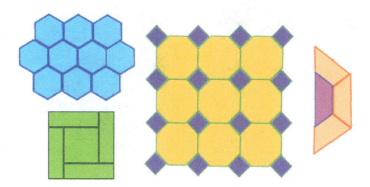

In any tiling, the key to making the figures fit tightly lies in the interior angles of the polygons used. The polygons must be aligned in a way so that the sum of the adjacent angles is 360°. Here are a few examples.

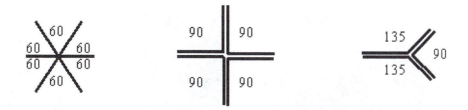

Why must the sum of the interior angles be 360°? In order to close all the gaps, imagine rotating around the point of intersection. Since there are 360° in a circle, we need the sum of the angles at that intersection to be 360°.

If the word "tiling" seems inappropriate to you, think of a kitchen floor or a stone patio, which are often covered in square tiles.

Tilings do not have to be made up of regular polygons. In many home stores, we can purchase bricks or oddly shaped stones that can be turned in different directions and form a tiling when placed together. These tilings can be used to make a unique sidewalk. Notice in the sidewalk pictured below on the left, all the stones are the same shape—just turned in different directions.

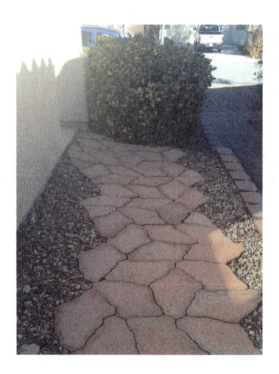
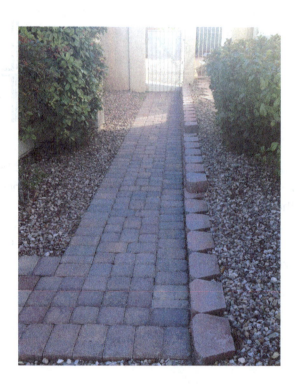

Section 5.3 Exercises

1. Classify each of the following as simple or non-simple.

   a.    b.    c.

2. How many sides does a hexagon have?

3. What is the name for a 15-sided polygon?

4. Classify the following triangles as equilateral, isosceles, or scalene.

   a.   b.   c.

5. Classify the following triangles as right, acute, or obtuse.

   a.   b.   c.

6. Indicate whether the following statements are true or false. If false, correct the statement.

   a. In a parallelogram, both pairs of opposite sides are parallel.

   b. All squares are rectangles.

   c. All rectangles are squares.

   d. All rectangles are parallelograms.

   e. All squares are parallelograms.

   f. In a rectangle, exactly two angles measure 90°.

7. Consider the following street signs.

   a. Which of the signs is not shaped like a polygon?

   b. Which of the signs are shaped like regular polygons?

   c. What is the name of the shape used for the STOP sign?

8. What is the measure of each of the interior angles for a regular pentagon?

9. What is the measure of each of the interior angles for a regular hectagon (100 sides)?

10. Can you make a tiling with only regular pentagons? Why or why not?

11. A regular heptagon has seven sides. Can you list a practical use for something shaped like a regular heptagon?

12. The Pentagon is the headquarters for the U.S. Department of Defense. Why is the name "Pentagon" appropriate?

Answers to Section 5.3 Exercises

1. a. simple   b. non-simple   c. simple

2. six

3. pentadecagon

4. a. isosceles   b. equilateral   c. scalene

5. a. acute   b. acute   c. right

6. a. true
    b. true
    c. false—possible rewrite: Some rectangles are squares.
    d. true
    e. true
    f. false—possible rewrite: In a rectangle, all four angles measure 90°.

7. a. Railroad crossing   b. stop ahead, stop, and yield   c. octagon

8. 108°

9. 176.4°

10. No. 108 is not a factor of 360. Three 108° angles add up to 324°, and four would be 432°.

11. Answers may vary, but one possible use is a daily pill container.

5.4 Redecorating Tips (Perimeter & Area)

12. The building is shaped like a pentagon.

We have all seen squares and rectangles, and remember some basic facts from previous studies. A **rectangle** is a flat, four-sided geometric figure in which both pairs of opposite sides are parallel and equal, and all four angles are right angles. A **square** is a rectangle in which all four sides are of equal length.

Perimeter & Area

Simply put, the **perimeter** of any flat geometric figure is defined as the distance around the figure. Although we may be presented with different formulas that get used for different figures, the perimeter of a given figure will still always be the distance around it.

The **area** of a flat geometric figure is defined as the amount of surface the figure covers. For rectangle and squares, this can be viewed as the number of square units that can be enclosed by the figure. In fact, for that very reason, area is always given in square units. For example, imagine a square that measures 1 inch on each side. How many of those 1-inch squares—also called "square inches"—can fit inside a rectangle that measures 2 inches wide by 5 inches long? If you're not sure, draw a picture.

IMPORTANT: When we are computing areas and perimeters we need to pay attention to the units of measure. If no unit is stated, then only numeric answers should be given. If, however, a unit is stated — such as centimeter, foot, or inch — we must include the appropriate unit with our answers.

Linear units should accompany the linear measures of perimeter. Areas should be stated in square units.

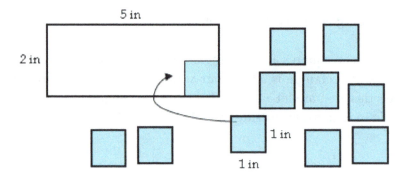

EXAMPLE 1: A rectangular room measures 3 yards by 4 yards. How many square yards of carpet are needed to cover the floor? Also, how many yards of baseboard trim will be needed to go around the base of the room? Assume the door to the room is 1 yard wide.

For the carpet, we need the area of the floor.

3 yd × 4 yd = 12 yd²

We need 12 square yards of carpet.

For the baseboard trim, we need to find the perimeter of the room, less the width of the door.

3 yd + 4 yd + 3 yd + 4 yd − 1 yd (for the doorway) = 13 yd

We need 13 yards of baseboard trim.

Notice how we were able to perform the previous example without the use of formulas. In the case of a perimeter, whether we have a rectangle or not, all we need to do is add up the

lengths of all the sides of the figure. When dealing with areas, we have already seen that the area of a rectangle is the product of the length and width. In formula form, if we use l for the length and w for the width, the formula is $A = lw$. Keep in mind, if we understand the origin of the formula, there is no need to memorize it.

Parallelograms

A **parallelogram** is a four-sided figure in which both pairs of opposite sides are equal in length and parallel, but the angles are not necessarily right angles.

Just like any flat geometric figure, the perimeter of a parallelogram is the distance around it. To find the area, we first need to identify the lengths of the base and height of the parallelogram. The **base**, b, of the parallelogram is the length of the bottom side. Actually, we can use any side, but, for simplicity, let's stick with the bottom. The **height**, h, of a parallelogram is the perpendicular distance from the base to the opposite side. This is usually represented by a dashed line drawn perpendicular to the base. Do not confuse the height with the length of a side. That is why we use a dashed line instead of a solid line.

To determine the area of a parallelogram, imagine cutting off the triangular region on the right side and moving it to the left side as follows.

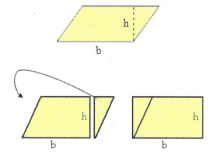

The resultant figure would be a rectangle of dimensions b and h. Thus, the area of the parallelogram would be the product of the base and the height, or $A = bh$.

EXAMPLE 2: Find the perimeter and area of the following parallelogram.

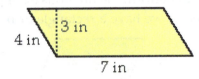

The perimeter, P, is the sum of the four side lengths. Remember, the 3 inch measure is the height, not a side length.

P = 4 in + 7 in + 4 in + 7 in = 22 in

The area is the product of the base and the height. In this case, the base is 7 inches and the height is 3 inches.

A = (7 in)(3 in) = 21 in²

Triangles

We all know what a triangle looks like, and once again, the perimeter of a triangle is just the sum of the three side lengths. For the area, we just need to recognize a triangle is literally half of a parallelogram, when we cut it along the diagonal.

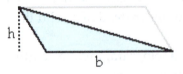

The height and base lengths are the same as with a parallelogram, but since we literally have half a parallelogram, the area of a triangle is half that of the parallelogram with the same base and height. That is, A = (1/2)(b)(h).

EXAMPLE 3: Find the perimeter and area of the following triangle.

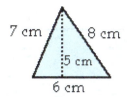

The perimeter: P = 7 cm + 8 cm + 6 cm = 21 cm

The area: A = (1/2)(6 cm)(5 cm) = 15 cm^2

Composite Figures

A **composite figure** is made up of two or more smaller figures. Like always, the perimeter of such a figure is still the distance around the figure. If the composite contains only right angles, we cut the figure into squares and rectangles, find the areas of the smaller pieces, and add them together to get the total area, which we will write as A_{Total}. Do remember, though, composite figures can also be made of other shapes, as well.

EXAMPLE 4: Find the area of the following composite figure.

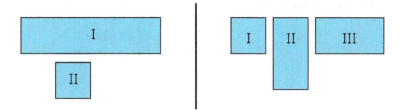

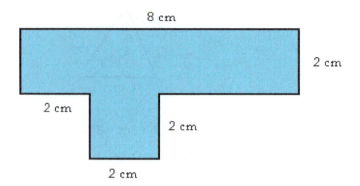

8 cm

2 cm

2 cm

2 cm

2 cm

2 cm

The figure can be viewed as a 2 cm by 8 cm rectangle on top of a 2 cm by 2 cm square, or as a 2 cm by 2 cm square next to two rectangles.

Below is the total for the area of the figures split on the left, and you can verify the area of the split on the right.

The area of the composite figure is the sum of the rectangle, Region I, and the square, Region II.

$$A_{Total} = A_{Rectangle} + A_{Square} = 16 \text{ cm}^2 + 4 \text{ cm}^2 = 20 \text{ cm}^2$$

It is important to note that, for the perimeter, we sum up *only* the sides that form the border of the composite figure. We should not try to break apart a figure to find the perimeter, but we may need to use some critical thinking skills to determine the lengths of unidentified sides.

EXAMPLE 5: Find the area and perimeter of the following figure.

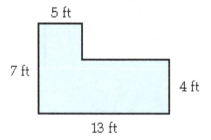

5 ft

7 ft

4 ft

13 ft

Before we can accurately compute the area and perimeter, we must first determine the lengths of the two unlabeled sides. For the missing horizontal measure, we need to notice the length of the bottom is 13 feet, and the length of the top-most side is 5 feet. That means the missing horizontal measure must be 8 feet. Likewise, the missing vertical measure can be found to be 3 feet.

Next, for the area, we can imagine the composite as a 5-foot by 3-foot rectangle on top of a 13-foot by 4-foot rectangle. Thus, the total area, A = (5 ft)(3 ft) + (13 ft)(4 ft) = 15 ft² + 52 ft² = 67 ft².

For the perimeter, we add together the lengths of all six sides. Be careful not to forget the lengths of the unlabeled sides. Starting at the top and going clockwise, we find the perimeter, P = 5 ft + 3 ft + 8 ft + 4 ft + 13 ft + 7 ft = 40 ft.

EXAMPLE 6: Find the area and perimeter of the following figure.

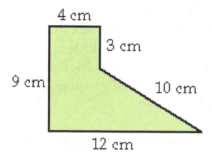

For the area, we should imagine the figure as a 4 cm by 9 cm rectangle with a triangular piece next to it. BE CAREFUL! The base length of that triangle part is not 12 cm; it is only 8 cm. Also, how do we find the height of the triangular piece?

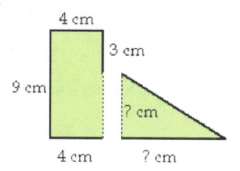

As we have seen, since the length of the base for the whole composite figure is 12 cm, and 4 cm of that is composed of the rectangle, the triangle-shaped region has a base of 8 cm. For the height of the triangular region, we see the entire height of the composite figure is 9 cm, and the triangle height is 3 cm lass than that. So, the height of the triangle is 6 cm.

Now, taking the entire composite figure in mind,

$$A_{Total} = A_{Rectangle} + A_{Triangle}$$

$$A_{Total} = (4 \text{ cm})(9 \text{ cm}) + (1/2)(8 \text{ cm})(6 \text{ cm}) = 36 \text{ cm}^2 + 24 \text{ cm}^2 = 60 \text{ cm}^2.$$

For the perimeter, start at the top and go clockwise around the figure. $P = 4 \text{ cm} + 3 \text{ cm} + 10 \text{ cm} + 12 \text{ cm} + 9 \text{ cm} = 38 \text{ cm}$.

A Few Final Words

Linear distances— such as feet (ft), inches (in), and miles (mi)—are used quite frequently in our daily lives. Likewise, areas—such as square feet (ft²), square inches (in²) and square miles (mi²)—are pretty common, too. Additionally, there is another relatively common measure: **Volume**. Refrigerators have capacity stated in cubic feet (ft³), we pour concrete in cubic yards (yd³), and we buy gallons of milk, water, and gasoline. Volumes point to the third dimension in our three-dimensional world.

Similar to how the area of a rectangle was found as the length times the width, the volume of a **right rectangular parallelepiped** (i.e., a **box**) is found by multiplying the length by the width by the height, $V_{box} = l \times w \times h$.

Be sure to pay attention to the unit of measure. Many volumes are stated in imperial measures, like gallons and pints, or in metric measures, like liters and milliliters. If, however, a volume is computed from the product of linear dimensions, it needs to have a cubic unit. That is, if we compute the volume by finding the product of a length, width, and height that were all measured in centimeters, the final unit of measure on the volume would be cm³. This is just like multiplying $a \times a \times a$ to get a^3, cm × cm × cm is cm³,

EXAMPLE 7: Find the volume of a box that has a length of 5 inches, a width of 8 inches, and a height of 4 inches.

$$5 \text{ in} \times 8 \text{ in} \times 4 \text{ in} = 160 \text{ in}^3$$

Beyond boxes, unfortunately, deriving the volume formulas for various three-dimensional figures can become quite complicated. Thus, in order to perform many of those calculations, we often have to resort to the memorization of formulas. If you are curious, you can see some of them at http://math.com/tables/geometry/volumes.htm.

Section 5.4 Exercises

1. Find the perimeter and area of a square that has a side length 5 inches.

2. Find the perimeter and area of a square with a side length of 3 miles.

3. Find the perimeter and area of a rectangle with a length of 2 meters and a width of 3 meters.

4. Find the perimeter and area of a rectangle with a length of 2 feet and a width of 9 feet.

5. Find the perimeter and area of a rectangle with a length of 5 cm and a diagonal of 13 cm.

6. Find the perimeter and area of a rectangle with a length of 6 ft and a diagonal of 10 ft.

7. Find the perimeter and area of the following parallelogram.

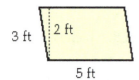

8. Find the perimeter and area of the following parallelogram.

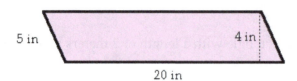

9. Find the area of a triangle with a base of 12 inches and a height of 3 inches.

10. Find the area of a triangle with a base of 7 meters and a height of 4 meters.

11. Find the area and perimeter of the following triangle.

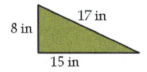

12. Find the area and perimeter of the following triangle.

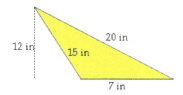

13. A rectangular room measures 10 feet by 12 feet. How many square feet of tile are needed to cover the floor?

14. John's kitchen is rectangular and measures 11 feet wide by 13 feet long, and he has a 3-ft by 4-ft island in the middle of the kitchen. Assuming no tiles are placed under the island, how many square feet of tile will John need to tile the floor?

15. A skirt is to be placed around a rectangular table that is 6 feet long and 2 feet wide. How long does the table skirt need to be to fit around the entire table (without overlapping)?

16. Find the area and perimeter of the following figure.

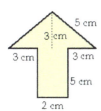

17. Find the area and perimeter of the following figure.

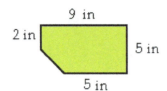

18. Find the area and perimeter of the following figure.

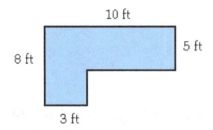

19. Find the area and perimeter of the following figure.

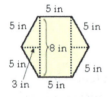

20. Find the area and perimeter of the following figure.

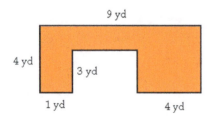

21. Find the area of the shaded region in the following figure.

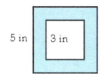

22. Draw two rectangles that have the same area, but different perimeters.

23. Draw two rectangles that have the same perimeter, but different areas.

24. If the area of a rectangle is 68 m² and the width is 4 m, what is the length?

25. If the area of a rectangle is 36 in² and the length is 12 in, what is the width?

26. Draw two different triangles that have the same area but different perimeters.

27. Draw two different triangles that have the same perimeter but different areas.

28. If the area of a triangle is 42 cm² and the height is 7 cm, what is the length of the base?

29. If the area of a triangle is 12 ft² and the base length is 3 ft, what is the height?

30. What is the volume of a box with a length of 3 cm, a width of 9 cm, and a height of 5 cm?

Answers to Section 5.4 Exercises

1. P = 20 in, A = 25 in²

2. P = 12 mi, A = 9 mi²

3. P = 10 m, A = 6 m²

4. P = 22 ft, A = 18 ft²

5. P = 34 cm, A = 60 cm²

6. P = 28 ft, A = 48 ft²

7. P = 16 ft, A = 10 ft²

8. P = 50 in, A = 80 in²

9. A = 18 in²

10. A = 14 m²

11. A = 60 in², P = 40 in

12. A = 42 in², P = 42 in

13. 120 sq ft of tile are needed.

14. 131 sq ft of tile are needed.

15. The skirt must be 16 ft long.

16. A = 22 cm², P = 28 cm

17. A = 39 in², P = 26 in

18. A = 59 ft², P = 36 ft

19. A = 64 in², P = 30 in

20. A = 24 yd², P = 32 yd

21. 16 in²

22. Answers can vary. One possible pair is a 6 × 2 rectangle and a 3 × 4 rectangle.

23. Answers can vary. One possible pair is a 5 × 2 rectangle and a 3 × 4 rectangle.

24. The length would be 17 meters.

25. The width would be 3 inches.

26. Answers can vary. For example, a triangle with a base of 6 in and a height of 2 in has the same area as a triangle with a base of 4 in and a height of 3 in. If you draw them carefully you will see the first one has a greater perimeter.

27. Answers can vary. For example, a triangle with side lengths of 15 in, 20 in, and 25 in has a perimeter of 60 in and an area of 150 in², while a triangle with side lengths of 10 in, 24 in, and 26 in has a perimeter of 60 in and an area of 120 in².

28. The base length is 12 cm.

29. The height is 8 ft.

30. 135 cm³

5.5 Spinning Wheels (Circles)

Circles

All of us can identify a circle by sight, so we will focus on a few certain properties of circles. The **diameter** of a circle is the distance across a circle as measured through the circle's center, and is indicated with the letter d. Instead of the word perimeter, the distance around a circle is called its **circumference**, and is indicated with the letter C.

Experiment

Select three circular objects, and use a piece of string and a metric ruler to measure the circumference and diameter of each. Either print this sheet or make a list that looks like the one below. In each case, compute C/d (rounded to the nearest hundredth).

Object	C	d	C/d
_____	_____ cm	_____ cm	_____
_____	_____ cm	_____ cm	_____
_____	_____ cm	_____ cm	_____

Pi & r

For each of the above circular objects, you should have found C/d to be close to 3.14. This value is represented by the lowercase Greek letter pi, π. If you did everything accurately in the above experiment, you should have found the value of C/d to be really close to pi for each one. Also, the **radius**, r, is defined as half of the circle's diameter.

EXAMPLE 1: Find the radius of a circle with diameter 12 centimeters.

The radius is half the diameter, so r = 6 cm.

EXAMPLE 2: Find the diameter of a circle with a radius of 9 feet.

The diameter is twice the radius, so d = 18 ft.

Area & Circumference Formulas

Using *r* for the radius, and *d* for the diameter, the area and circumference of a circle are defined as:

Area, A = πr²

Circumference, C = πd. Or, since d = 2r, we often see this as C = 2πr.

Also, to approximate the calculations, it is customary to use π = 3.14. Use these formulas to find the area of the three circles in your experiment.

EXAMPLE 3: Find the area and circumference for a circle with a radius of 3 inches. Be sure to label your answers. Use π = 3.14, and round your answers to the nearest hundredth.

Area = A$_C$ = πr² = (3.14)(3 in.)² = 28.26 in²

Circumference = C = 2πr = 2(3.14)(3 in) = 18.84 in

EXAMPLE 4: Find the area and circumference of the following circle. Be sure to label your answers. Use π = 3.14, and round your answers to the nearest tenth.

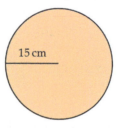

15 cm

Area = (3.14)(15 cm)² = 706.5 cm²

Circumference = 2(3.14)(15 cm) = 94.2 cm

Section 5.5 Exercises

For Exercises #1 through #17, be sure to label your answers. Use π = 3.14, and round your answers to the nearest tenth.

1. What is the radius length of a circle with diameter 12 inches?

2. What is the radius length of a circle with a diameter of 25 feet?

3. Find the area and circumference of a circle with radius 10 centimeters.

4. Find the area and circumference of a circle with radius 20 centimeters.

5. Find the area and circumference of a circle with radius 12 meters.

6. Find the area and circumference of a circle with radius 15 millimeters.

7. Find the area and circumference of a circle with radius 0.25 inches.

8. Find the area and circumference of a circle with radius 0.75 miles.

9. Find the area and circumference of the following figures.
   a.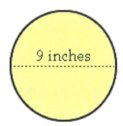
   b.

10. Find the area of the following figures.
    a.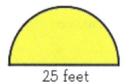
    b.

11. Find the area of the shaded region in the following figure.

5 m 5 m

12. Imagine the surface of the earth to be circular at the equator. If a steel band with a circumference of 10 feet more than the circumference of the earth is placed around the equator and spaced equidistant away from the surface, which of the following animals could walk comfortably under the band?
   a. A mouse.   b. A cat.   c. A dog.   d. An elephant.   e. A giraffe.

13. A fly lands on the tip of the minute hand of a circular clock. If the min
   long, how far has the fly traveled if it remains on the minute hand for

14. A table skirt is to be wrapped around a circular table with a diameter of 6 feet. How long does the skirt need to be?

15. If you double the circumference of a circle, by how much has the area changed?

16. If the area of a circle is doubled, by how much has the circumference changed?

17. A pizza company offers a single large 16-inch (diameter) pizza or a pair of medium 12-inch pizzas for the same price. Which option gives you more pizza?

Answers to Section 5.5 Exercises

1. 6 in

2. 12.5 ft

3. $A = 314$ cm² & $C = 62.8$ cm

4. $A = 1256$ cm² & $C = 125.6$ cm

5. $A = 452.2$ m² & $C = 75.4$ m

6. $A = 706.5$ mm² & $C = 94.2$ mm

7. $A = 0.2$ in² & $C = 1.6$ in

8. $A = 1.8$ mi² & $C = 4.7$ mi

9. a. $A = 63.6$ in² & $C = 28.3$ in

   b. $A = 254.3$ ft² & $C = 56.5$ ft

10. a. $A = 245.3$ ft²

    b. $A = 490.6$ ft²

11. $A_{Total} = A_{Big\ Circle} - A_{Small\ Circle} = (3.14)(10\ m)^2 - (3.14)(5\ m)^2 = 235.5$ m²

12. To answer this, find the difference in the radii of the two circles. So, if r_1 is the radius of the earth, and r_2 is the radius of the steel band, we need to find $r_2 - r_1$. $2\pi r_2 = 2\pi r_1 + 10$. Subtract $2\pi r_1$ from both sides, and then factor out and divide by the 2π. You will find $r_2 - r_1 = 10/2\pi = 10/6.28 = 1.59$ ft, which is about 19 inches. So, with just 10 feet added to the circumference, the radius has increased by a little more than a foot and a half. That means the mouse and the cat can comfortably walk under the band. The dog could fit if it is a small-to-medium sized dog. A large dog could make it, but not comfortably. The elephant and giraffe won't fit.

13. In 12 minutes, the hand has traveled $12/60 = 1/5$ of an hour. Thus, the tip of the minute hand has traveled $1/5$ of the circumference of the circle or radius 8 in. $C = 2(3.14)(8\ in) = 50.24$ in. $50.24/5 = 10.048$ in. The fly has traveled a little over 10 inches.

14. 18.84 feet. By the way, the 0.84 is $84/100$ of a foot, which is just over 10 inches. So, the skirt would need to be at least 18 feet 10 inches long.

15. In the formula $C = 2\pi r$, only the r can change. So, to double the circumference, the radius must be doubled. In the area formula $A = \pi r^2$, the radius is squared: $(2r)^2 = 4r^2$. Thus, the new area would be $\pi(4r^2)$, which would be written $4\pi r^2$. So, the area has been quadrupled.

16. In the formula $A = \pi r^2$, only the r can change. So, to double the area, the radius must be increased by a factor of $\sqrt{2}$. Thus, the circumference is also increased by a factor of $\sqrt{2}$.

17. The large pizza has an area of 201 square inches, and the pair of medium pizzas has an area of 226 square inches. You get more with the two medium pizzas.

Credits

1. "Compass," http://pixabay.com/en/dividers-circle-compasses-154067/. Copyright in the Public Domain.
2. Copyright © Wikimedia Foundation, Inc. (CC BY-SA 3.0) at http://en.wikipedia.org/wiki/Pythagoras.
3. Copyright © Wellcome Images (CC by 4.0) at http://wellcomeimages.org/indexplus/image/V0004826.html.
4. Copyright © Wikimedia Foundation, Inc. (CC BY-SA 3.0) at http://en.wikipedia.org/wiki/Euclid.
5. Copyright © Wikimedia Foundation, Inc. (CC BY-SA 3.0) at http://en.wikipedia.org/wiki/Euclid%27s_Elements.
6. "Eukleides of Alexandria," http://commons.wikimedia.org/wiki/File:Euklid-von-Alexandria_1.jpg. Copyright in the Public Domain.
7. "Portrait of Hypatia," http://commons.wikimedia.org/wiki/File:Hypatia_portrait.png. Copyright in the Public Domain.
8. Copyright © Wikimedia Foundation, Inc. (CC BY-SA 3.0) at http://en.wikipedia.org/wiki/Hypatia.
9. "Cartoon Woman," http://pixabay.com/en/woman-lady-cartoon-nose-big-nose-312040/. Copyright in the Public Domain.
10. "Cartoon Tree," http://pixabay.com/en/tree-forest-trunk-nature-leaves-576847/. Copyright in the Public Domain.
11. "Cartoon Woman," http://pixabay.com/en/woman-lady-cartoon-nose-big-nose-312040/. Copyright in the Public Domain.
12. "Cartoon Tree," http://pixabay.com/en/tree-forest-trunk-nature-leaves-576847/. Copyright in the Public Domain.
13. "Ladder," http://pixabay.com/en/ladder-upload-get-achievement-434523/. Copyright in the Public Domain.
14. "Baseball Field," http://pixabay.com/en/baseball-field-sports-ball-game-192400/. Copyright in the Public Domain.
15. "Cartoon TV," http://pixabay.com/en/flat-screen-tv-television-screen-32307/. Copyright in the Public Domain.
16. "Cartoon Man," http://pixabay.com/en/man-character-person-stand-look-37334/. Copyright in the Public Domain.
17. "Cartoon Lamp," http://pixabay.com/en/streetlight-light-street-urban-157598/. Copyright in the Public Domain.
18. "Patio Furniture," http://pixabay.com/en/teak-teak-patio-furniture-172642/. Copyright in the Public Domain.
19. Copyright © David B. Gleason (CC BY-SA 2.0) at http://commons.wikimedia.org/wiki/File:The_Pentagon_January_2008.jpg?fastcci_from=410440.
20. "Clock Face," http://pixabay.com/static/uploads/photo/2014/09/08/21/37/clock-439592_640.png. Copyright in the Public Domain.

6 VOTING AND APPORTIONMENT

Voting is the foundation of any democracy. However, how we vote and what we vote on varies greatly. An Ancient Greek politician, Cleisthenes introduced one of the earliest forms of democracy in 508 BC. With a very large government representing millions of people, there was a voting process that can be thought of as a negative election—what we might call *dis*-approval voting—where, every year, voters were asked to cast a vote for the politician they most wished to *exile* for 10 years. Votes were written on broken pots known as **ostraka** (the origin of the word "ostracize"). If no politician received more than 6,000 votes, they all remained. If anyone received more than 6,000 votes, the politician with the largest number of votes was exiled. Requiring someone to have over 6,000 votes before being ostracized was done to ensure politicians would only be exiled if they were unpopular with a large number of people.

6.1 On the Shoulders of Giants (Biographies & Historical References)

Elections can get very nasty, but few were worse than the Presidential election of 1800. During that campaign, Federalist newspapers claimed the election of Thomas Jefferson would cause the "teaching of murder, robbery, rape, adultery, and incest." That election also saw Alexander Hamilton attempt to sabotage the efforts of his own Federalist party because of a personal dislike of its candidate, the incumbent President John Adams. In the same election, Hamilton ended up breaking a tie between Jefferson and Aaron Burr. Because he detested Burr, Hamilton lobbied heavily against him.

The election of a President depends greatly on the population of the individual states within the United States. Despite his questionable influence in a couple elections, Alexander Hamilton devised a system of determining the number of votes awarded to each state. Although originally vetoed by President George Washington (the very first exercise of the veto power by the President of the United States), the Hamilton Method of Apportionment was later adopted by Congress and used for many years.

In the United States, many modern-day voters in Presidential elections often think they are voting directly for a specific candidate. In actuality, they are indicating how they want an official, called an elector, to vote. Additionally, even if the majority of a state's voters prefer a specific candidate, the identified electors in the Electoral College can actually vote for a different candidate.

The Presidential Election of 1800

In early Presidential elections, electors voted for two candidates, and the candidate with the most votes was named President, while the one with the second most votes became the Vice President. A typical strategy by a political party would see that one elector would abstain from casting a second vote, thereby allowing their intended Presidential candidate to have one more vote than their intended Vice Presidential candidate.

The Presidential Election of 1800 was one of the nastiest political campaigns ever. What made things more interesting was that it pitted the sitting President, Federalist John Adams, against the sitting Vice President, Democratic-Republican Thomas Jefferson. The Federalists also had Charles Cotesworth Pinckney as a candidate, and the Democratic-Republican Party added Aaron Burr to the slate of candidates.

John Adams faced substantial re-election opposition within his own party. Alexander Hamilton schemed to have Pinckney, the implied Vice Presidential candidate, receive more electoral votes and, thus, become President. Unfortunately for the Federalists, the Democratic-Republican team of Jefferson and Burr got more votes. And, unfortunately for the Democratic-Republicans, Jefferson and Burr each got the *same number* of votes. Since no single candidate had a majority, the election was turned over to the House of Representatives, which had a large number of Federalist members.

The House deliberated for seven days and voted 36 times. Many Federalist members of the House detested Jefferson and, hence, supported Burr for President. The Democratic-Republican members, on the other hand, wanted Jefferson to be President. Alexander Hamilton hated Aaron Burr much more than he hated Thomas Jefferson, and through his considerable influence (and back-stabbing) he succeeded in getting some Representatives to switch their votes and, on the 36th ballot, Jefferson was selected President, making Aaron Burr the Vice-President.

Alexander Hamilton

Alexander Hamilton (1755–1804) was born in Charlestown, the capital of Nevis in the British West Indies and was raised in the Caribbean. He was one of the Founding Fathers of the United States, the first Secretary of the Treasury, and served as a military officer and confidant to George Washington during the American Revolutionary War. He was one of America's first lawyers, wrote the majority of the *Federalist Papers*, which were, and continue to be, a leading source for interpretation of the U.S. Constitution, and contributed to the development of many theories related to economics and political science.

After coming to the colonies, Hamilton attended King's College (now Columbia University) in New York City. At the start of the American Revolutionary War, he studied military history and tactics, raised the New York Provincial Company of Artillery, and became a captain in the Continental Army. When he reached the rank of Lieutenant Colonel, Hamilton joined George Washington's staff in March of 1777. He served Washington for four years and was involved in a wide variety of high-level duties, including intelligence, diplomacy, and negotiation with senior army officers. The importance of these duties makes a statement about the high level of confidence Washington had in him.

Hamilton's intense rivalry with Burr would provide a tragic ending to Hamilton's life. Following an exchange of nasty letters, and despite the attempts of friends to avert a confrontation, a duel was scheduled for July 11, 1804, along the west bank of the Hudson River in Weehawken, New Jersey. Ironically, this was the same dueling site where Hamilton's eldest son, Philip, was killed three years earlier. At dawn on the scheduled date, then Vice President Aaron Burr shot Hamilton, and the injuries sustained resulted in Hamilton's death the next afternoon.

The Electoral College

While the total number of votes for each candidate is made public, the popular vote is not the basis for determining the winner of a presidential election in the United States. The United States Electoral College is an example of an "indirect election," because rather than directly voting for the President, United States citizens cast votes that direct the 538 members of the college how to elect the President.

Although there is nothing in the United States Constitution requiring the winner of the popular vote within each state to be given *all* of the electoral votes from that state, this has been the tradition since the founding of the country. At present, only Maine and Nebraska split their electoral votes among the candidates, and these states give their votes to the candidate that wins each congressional district.

The number of electoral votes for a specific state is based on population, and corresponds to the number of Senators and Representatives the state has in Congress. The states with the smallest populations have two Senators and one Representative and are assigned three electoral votes. California, which currently has the largest population of all the states, has two Senators and 53 Representatives, for a total of 55 electoral votes. The following map shows the Electoral College distribution as of 2012, which is based on the 2010 United States Census.

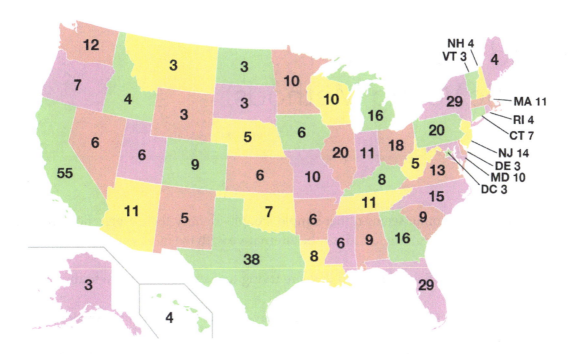

A majority of the 538 electoral votes is needed to win the election. So, since half-votes are not allowed, a candidate must obtain 270 electoral votes win. There have actually been three elections in which the winner, via the Electoral College, did not receive the highest number of popular votes: 1876, 1888, and 2000.

The constitutional theory behind the indirect election of the president of the United States is, while the members of Congress are elected by popular vote, the President is considered to be the Executive of a Federation of Independent States. Therefore, each state selects its preferred candidate and gives its electoral votes to that candidate.

Critics of the Electoral College argue it is inherently undemocratic and gives certain swing states disproportionate clout in selecting the President. Proponents argue that the Electoral College is an important and distinguishing feature of federalism in the United States and protects the rights of smaller states. Numerous constitutional amendments have been introduced in Congress seeking a replacement of the Electoral College with a direct popular vote; however, no proposal has ever been passed.

6.2 And the Winner Is… (Basic Voting Methods)

On the surface, voting seems rather simple. People identify their preference, and the votes are counted. But is it fair to say the alternative with the most votes is the winner?

Suppose Alice, Bob, Carol, and Don are trying to decide where to go to dinner.

- Alice wants Chinese food, is willing to have Italian food, but hates Mexican food.

- Bob wants Mexican food, is willing to have Italian food, but hates Chinese food.

- Carol wants Mexican food, is willing to have Italian food, but hates Chinese food.

- Don loves Italian food, is willing to have Chinese food, but hates Mexican food.

If everyone stated only his or her main preference, Mexican food would get two votes, and each of the other two would get a single vote. Thus, the foursome would head to a Mexican restaurant. However, is that really the best choice?

Think about it; half of the group hates Mexican food. And, since all four people are willing to have Italian food, perhaps it is the best choice. That way, no one in the group would be unhappy. Counting only everyone's favorite choice takes care of the likes, but completely ignores the dislikes. When we take the dislikes into account, the situation can turn out very differently.

Preference Lists

When there are several alternatives to choose from in a situation where voting will take place, one method is to have each voter **rank** the alternatives. When this ranking is created, it is called a **preference list**.

Let's say ten people will rank their favorite flavors of ice cream, with the choices being Chocolate, Vanilla, and Strawberry.

					Specific Voters					
	#1	#2	#3	#4	#5	#6	#7	#8	#9	#10
First Choice	C	S	V	S	S	C	V	S	C	S
Second Choice	S	V	S	C	C	S	S	C	S	C
Third Choice	V	C	C	V	V	V	C	V	V	V

As we look over the choices that were made, we can combine identical rankings and summarize this group of ballots into the following **preference list**.

		Number of Voters		
	4	3	2	1
First Choice	S	C	V	S
Second Choice	C	S	S	V
Third Choice	V	V	C	C

From the preference list, we see four people ranked the flavors in the order S, C, V; three people ranked the flavors C, S, V; two people ranked the flavors V, S, C; and one person ranked the flavors S, V, C.

Also, pay attention to the top rows in each of the two previous tables. The first one is an indication of how each voter indicated his/her preferences, and the second table represents a tally of all the votes. These are subtle but important differences.

EXAMPLE 1: Use the information from the ice cream preference list (above) to answer the following:

a. How many voters had Strawberry as their favorite flavor?

b. How many people voted?

c. How many voters indicated Vanilla was their least favorite flavor?

Answers

a. 5   b. 10   c. 7

Preference Voting Methods

There are many different ways to determine the winner in a situation involving voting where a voter ranks the candidates, and we will investigate several of them in this section.

Plurality

Many of us are familiar with Presidential elections in the United States, which provide an example of the first type of voting that we will discuss. The method is very simple, and it is known as the **Plurality Method**.

While we will use preference list ballots as we study each different method of voting, in the Plurality Method, only the first-place votes are considered. **The candidate with the most first-place votes is the winner.**

> ### Plurality Voting Method
>
> ### The candidate or choice with the most first-place votes is the winner.

EXAMPLE 2: Use the provided preference list and the Plurality Voting Method to determine the winner.

	Number of Voters			
Rank	5	3	3	2
First Choice	C	B	B	D
Second Choice	D	D	A	A
Third Choice	A	A	D	C
Fourth Choice	B	C	C	B

In this preference list, Candidate A has zero first-place votes, Candidate B has six first-place votes, Candidate C has five first-place votes, and Candidate D has 2 first-place votes.

Since Candidate B has the most first-place votes, using the Plurality Method, Candidate B is the winner.

Borda Count

Named after a French military officer and mathematician, Jean-Charles de Borda, the **Borda Count Voting Method** assigns point values to each position within the preference list. In a Borda Count Method, if n candidates are ranked, a first-place ranking is worth n points, a second-place ranking is worth n-1 points, and so on, making a last-place ranking worth one point.

What we have referred to as the Borda Count method should really be considered a **Standard Borda Count** method. In a "standard" version of the Borda Count method, we always assign one point to the item or candidate in last place, two points for the one in the second-to-last place, and so on up the chain.

Unless stated otherwise, we will stick with a Standard Borda Count method. In a non-standard Borda Count method, different weights can be assigned to the places, as long as they are increasing in value for the places closer to first place. For example, if there are three candidates, 10 points can be assigned to a first place vote, 5 points to a second place vote, and 2 points for a third place vote. This strategy gives a distinct advantage to the candidate with the highest number of first place votes, especially if the weights are extreme.

If first place votes were worth 1,000 points, second place votes earned the candidate 2 points, and last place was worth 1 point, we would have something very close to a plurality system. The difference would be, in the event of a tie in the number of first place votes, the points earned from second and third place votes would serve as the tiebreaker. In Borda Count cases, the weight of each place must be identified *before* any voting begins. If someone were to see all the votes before the results get tabulated, it would be possible for that person to manipulate the results by declaring an extreme weight for first place. Keep in mind, if a Standard Borda Count Method is declared, specific point values may not be mentioned, as a first-place vote will have a value equal to the number of candidates.

Borda Count Voting Methods are used in a number of situations, including when polls are taken to rank collegiate sports teams.

Borda Count Voting Method

Assign point values to each position within a preference list. The candidate with the highest weighted total is the winner.

EXAMPLE 3: Use the provided preference list and the Standard Borda Count Voting Method to determine the winner.

Rank	Number of Voters			
	5	3	3	2
First Choice	C	B	B	D
Second Choice	D	D	A	A
Third Choice	A	A	D	C
Fourth Choice	B	C	C	B

In this election with four candidates, a first-place vote is worth four points, a second-place vote is worth three points, a third-place vote is worth two points, and a fourth-place vote is worth one point.

Be sure to read this preference list properly. Remember, the first column represents the rankings of **five voters.** That being the case, when we calculate the point totals for each candidate, the values of those rankings will get **multiplied by five.** The other columns will be treated similarly.

Let's take a look at the rankings for Candidate A.

From the first column, we can see five people ranked Candidate A as their third choice. Each third-place ranking is worth two points, so from this column, Candidate A will receive $5 \times 2 = 10$ points.

In the second column, we see three people ranked Candidate A as their third choice. Here again, each third-place ranking is worth two points, so from the second column, Candidate A will receive $3 \times 2 = 6$ points.

In the third column, we see three people ranked Candidate A as their second choice. Since each second-place ranking is worth three points, Candidate A will receive $3 \times 3 = 9$ points from this column.

Finally, in the fourth column, we see two people ranked Candidate A as their second choice. Again, each second-place ranking is worth three points, so Candidate A will receive $2 \times 3 = 6$ points from this column.

With that in mind, let's calculate the number of points for each candidate:

- Points for Candidate A: $(5 \times 2) + (3 \times 2) + (3 \times 3) + (2 \times 3) = 10 + 6 + 9 + 6 = 31$

- Points for Candidate B: $(5 \times 1) + (3 \times 4) + (3 \times 4) + (2 \times 1) = 5 + 12 + 12 + 2 = 31$

- Points for Candidate C: $(5 \times 4) + (3 \times 1) + (3 \times 1) + (2 \times 2) = 20 + 3 + 3 + 4 = 30$

- Points for Candidate D: $(5 \times 3) + (3 \times 3) + (3 \times 2) + (2 \times 4) = 15 + 9 + 6 + 8 = 38$

Candidate D has the highest weighted total. So, using the Standard Borda Count Method, Candidate D is the winner.

Hare System

Developed by Thomas Hare in 1861, and described by John Stuart Mill as "among the greatest improvements yet made in the theory of practice of government," the **Hare Voting System** eliminates the candidate who receives the least number of first-place votes and then recounts the ballots. This process is repeated until a single candidate has obtained the majority of first-place votes. This type of voting is used in elections in Australia, Ireland, and Northern Ireland and when determining the winners of the Academy Awards.

Hare Voting System

Eliminate the candidate with the least number of first-place votes, and recount the votes. Repeat this process until one candidate has a majority of first-place votes and is declared the winner.

The Hare System is also **known as the Instant Run-Off**, the **Single Transferable Vote (STV)**, or **Plurality with Elimination** voting method.

EXAMPLE 4: Use the provided preference list and the Hare Voting System to determine the winner.

	Number of Voters			
Rank	5	3	3	2
First Choice	C	B	B	D
Second Choice	D	D	A	A
Third Choice	A	A	D	C
Fourth Choice	B	C	C	B

There are 13 voters, and, thus, it will take seven first-place votes to be the winner. To use the Hare System, we look at the number of first-place votes and eliminate the candidate who has the fewest number of them.

- Candidate A has zero first-place votes.

- Candidate B has six first-place votes.

- Candidate C has five first-place votes.

- Candidate D has two first-place votes.

No single candidate has more than half of the first-place votes. So, continuing with the procedure, Candidate A has the fewest first-place votes and is eliminated.

With Candidate A deleted, the preference list now looks like:

	Number of Voters			
Rank	5	3	3	2
First Choice	C	B	B	D
Second Choice	D	D		
Third Choice			D	C
Fourth Choice	B	C	C	B

None of the first-place votes changed, but it is still a good idea to get in the habit of condensing the table after removing one of the candidates. That is, in each column, leave the votes in the same order, but decrease the number of options by one and remove any blanks in the table.

The condensed version of the table will look like:

	Number of Voters			
Rank	5	3	3	2
First Choice	C	B	B	D
Second Choice	D	D	D	C
Third Choice	B	C	C	B

Recounting the number of first-place votes for each remaining candidate, we have:

- Candidate B has six first-place votes.

- Candidate C has five first-place votes.

- Candidate D has two first-place votes.

Once again, no candidate has a majority. So, we keep going. This time, Candidate D has the fewest number of first-place votes and needs to be eliminated.

Deleting Candidate D from the list gives us the following:

	Number of Voters			
Rank	5	3	3	2
First Choice	C	B	B	
Second Choice				C
Third Choice	B	C	C	B

Now, because C has become the top choice of the voters represented in the fourth column, those two first-place votes belong to that candidate.

A condensed version of the preference list would look like:

	Number of Voters			
Rank	5	3	3	2
First Choice	C	B	B	C
Second Choice	B	C	C	B

With two candidates remaining, we recount the current number of first-place votes.

- Candidate B has six first-place votes

- Candidate C has seven first-place votes

Candidate C now has more than half of the first-place votes an is the winner.

Important Observation

So far, we have seen three different voting methods (plurality, Borda Count, and Hare) use the exact same preference list, and each method declared a different winner! If someone is allowed to view the ballots before a method is declared, knowledge of the different systems could be used to manipulate the results.

To take this a little further, we need to identify a voting system before any vote is cast. Voters need to know if they are to vote for a single candidate or multiple candidates. And, if they are voting for multiple candidates, they need to know whether or not their choices need to be ranked in a preference list.

Always identify a voting method before any vote is cast.

EXAMPLE 5: Use the provided preference list and the Hare System to determine the winner.

	Number of Voters			
Rank	7	5	3	2
First Choice	C	B	A	D
Second Choice	D	D	B	C
Third Choice	A	A	D	A
Fourth Choice	B	C	C	B

This preference list contains a total of 17 voters. In order for a candidate to have a majority and be declared the winner, he/she will need to have nine first-place votes.

As of now, the number of first-place votes for each candidate is as follows:

- Candidate A has three first-place votes.

- Candidate B has five first-place votes.

- Candidate C has seven first-place votes.

- Candidate D has two first-place votes.

Since no single candidate has the necessary nine first-place votes, the candidate with the lowest number of first-place votes will be eliminated.

Removing Candidate D, and condensing the preference list, we have:

Rank	Number of Voters			
	7	5	3	2
First Choice	C	B	A	C
Second Choice	A	A	B	A
Third Choice	B	C	C	B

The two voters who had Candidate D as their first choice now have Candidate C as their top choice.

The revised totals are:

- Candidate A has three first-place votes.

- Candidate B has five first-place votes.

- Candidate C has nine first-place votes.

Candidate C now has the required nine first-place votes and is declared the winner.

Pairwise Comparison

The **Pairwise Comparison Voting Method** examines only two candidates at a time, matching them up in head-to-head comparisons. The winning candidate from each head-to-head matchup gets one point. If there is a tie between two candidates, each candidate receives half a point. Every possible comparison must be made and, in the end, the candidate with the highest number of points is declared the winner.

Pairwise comparison is a technique that is used in round robin tournaments, such as the opening round of the World Cup Soccer tournament.

> ### Pairwise Comparison Voting Method
>
> Examine two candidates at a time, in a head-to-head matchup, and assign one point to the winner. The candidate with the highest point total is the winner.

Let's look at an example with four candidates: A, B, C and D. Remember, we have to look at every possible comparison, so we will need to find the winner of each of the following matchups.

- A against B

- A against C

- A against D

- B against C

- B against D

- C against D

EXAMPLE 6: Use the provided preference list and the Pairwise Comparison Voting Method to determine the winner.

	Number of Voters			
Rank	7	5	3	2
First Choice	C	B	A	D
Second Choice	D	D	B	C
Third Choice	A	A	D	A
Fourth Choice	B	C	C	B

For the matchup between Candidates A and B:

- From the first column, we see seven voters prefer A over B.

- From the next column, we see five voters prefer B over A.

- From the third column, three voters prefer A over B.

- From the last column, two voters have A ranked higher.

Thus, 12 of the 17 voters have A ranked higher than B. Therefore, from that matchup, A will get one point.

Once all the matchups are evaluated, we will see:

Matchup	Voter Preference	Assigned Points
A against B	12 of the 17 voters prefer A over B	**A** is given 1 point
A against C	9 of the 17 voters prefer C over A	**C** is given 1 point
A against D	14 of the 17 voters prefer D over A	**D** is given 1 point
B against C	9 of the 17 voters prefer C over B	**C** is given 1 point
B against D	9 of the 17 voters prefer D over B	**D** is given 1 point
C against D	10 of the 17 voters prefer D over C	**D** is given 1 point

Adding up all the points, we see:

- A has a total of one point.

- B has a total of zero points.

- C has a total of two points.

- D has a total of three points.

With the highest number of points, Candidate D is the winner.

When using the Pairwise Comparison Method, it is very important to make sure all possible head-to-head comparisons have been made. If we are familiar with counting techniques, we can use the formula for combinations to determine the number of comparisons. In the previous example, we needed to find the number of combinations of four candidates, taken two at a time.

$$_4C_2 = \frac{4!}{2!(4-2)!} = \frac{4\cdot3\cdot2\cdot1}{2\cdot1\cdot2\cdot1} = \frac{24}{4} = 6$$

Remember, this doesn't tell us which two candidates to compare; it just tells us *how many* comparisons are needed.

If we are not familiar with counting techniques, manually make sure each candidate is compared with every other candidate on the ballot. Do so in an organized fashion.

Start with the first candidate, and compare him with every candidate listed after him. Then move to the second candidate. Since the comparison between the first two candidates has already been done, start the comparison list for the second candidate with the third candidate in the list. Keep comparing each candidate with every candidate that comes after him. When we reach the last candidate, we are done.

Approval Voting

When the **Approval Voting Method** is used, each voter is allowed to cast a vote for as many candidates as he/she finds acceptable, and there is no limit to the number of candidates a voter can choose. The winner is the candidate who receives the highest number of votes.

Approval Voting Method

Each voter votes for as many candidates as he/she finds acceptable. The candidate with the most votes is the winner.

Since **no ranking of candidates is necessary**, we do not use a preference list. Instead, we use a table showing the votes cast. And, since this is not a preference list, the numbers under the heading of "Voters" do not indicate the number of people voting a certain way. They simply denote Voter #1, Voter #2, etc. Approval Voting is a **non-preferential voting method**.

EXAMPLE 7: Use the provided table of votes and the Approval Voting Method to determine the winner.

In this particular election, we have a total of eight voters voting for up to four candidates.

	Voters							
	#1	#2	#3	#4	#5	#6	#7	#8
Candidate A	✓		✓		✓	✓	✓	✓
Candidate B	✓	✓	✓	✓		✓	✓	✓
Candidate C		✓			✓			✓
Candidate D	✓		✓		✓	✓		

Counting the number of votes for each candidate, we have:

- Candidate A has a total of six votes.

- Candidate B has a total of seven votes.

- Candidate C has a total of three votes.

- Candidate D has a total of four votes.

So, using approval voting for this election, Candidate B is the winner.

It is also possible to use approval voting to declare several "winners." This is the approach used when selecting members of the National Baseball Hall of Fame. In order to qualify for induction into the Hall, a candidate needs to be selected by at least 70% of the voters. In terms of numbers, if there were 8 voters, a candidate would have to receive at least 5.6 votes to be elected. Since fractional votes are not permitted, it would take at least six votes for a candidate to be elected. In Example 7, both Candidate A and Candidate B were selected by at least 70% of the voters. Thus, if this ballot were for Hall of Fame voting, both of them would be declared "winners."

Summary of Basic Voting Methods

Preferential Methods

- Plurality: The candidate or choice with the most first-place votes is the winner.

- Borda Count: Assign point values to each position within a preference list. The candidate with the highest weighted total is the winner.

- Hare System: Eliminate the candidate with the least number of first-place votes, and recount the votes. Repeat this process until one candidate has a majority of first-place votes and is declared the winner.

- Pairwise Comparison: Examine two candidates at a time in a head-to-head matchup and assign one point to the winner or, in the event of a tie, a half point to each candidate. The candidate with the highest point total is the winner.

Non-Preferential Method

- Approval Voting: Each voter votes for as many candidates as he/she finds acceptable. The candidate with the most votes is the winner.

Section 6.2 Exercises

1. Match each voting method (a.–e.) with the phrase (i.–v.) that best describes it:

   a. Plurality Voting

   b. Borda Count

   c. Hare System

   d. Pairwise Comparison

   e. Approval Voting

   i. Points are given to candidates based upon their ranking in preference list ballots

   ii. Uses preference list ballots to delete candidates who receive the lowest number of first-place votes until one of the candidates has a majority of the votes

   iii. Candidates are matched against each other in head-to-head comparisons

   iv. The winner is the candidate that receives the highest number of first-place votes

   v. Each voter casts votes for every candidate that they find to be acceptable

2. If there are three candidates in an election, how many pairwise comparisons need to be made to determine the winner?

3. If there are six candidates in an election, how many pairwise comparisons need to be made to determine the winner?

4. These candidates received the following number of votes:

    Albert = 375, Benny = 210, Carol = 411, Donna = 189

    Using the Plurality Method, which candidate is the winner?

5. Using the provided preference list, answer the questions that follow it.

	Number of Voters			
Rank	3	2	2	1
First Choice	D	C	B	A
Second Choice	B	B	C	C
Third Choice	C	A	A	B
Fourth Choice	A	D	D	D

a. How many voters were there?

b. How many voters have ranked the candidates in the order D, B, C, A?

c. Who is the winner using Plurality Voting?

d. Who is the winner using the Borda Count Method?

e. Who is the winner using the Hare System?

f. Who is the winner using Pairwise Comparison?

6. Using the provided preference list, answer the questions that follow it.

	Number of Voters			
Rank	4	2	2	1
First Choice	B	C	A	A
Second Choice	D	A	C	B
Third Choice	C	D	B	C
Fourth Choice	A	B	D	D

a. How many voters were there?

b. How many voters have ranked the candidates in the order A, B, C, D?

c. Who is the winner using Plurality Voting?

d. Who is the winner using the Borda Count Method?

e. Who is the winner using the Hare System?

f. Who is the winner using Pairwise Comparison?

7. Using the provided preference list, answer the questions that follow it.

	Number of Voters			
Rank	5	3	2	1
First Choice	C	B	D	B
Second Choice	B	A	C	A
Third Choice	D	D	B	D
Fourth Choice	A	C	A	C

a. How many voters were there?

b. How many voters have ranked the candidates in the order C, B, D, A?

c. Who is the winner using Plurality Voting?

d. Who is the winner using the Borda Count Method?

e. Who is the winner using the Hare System?

f. Who is the winner using Pairwise Comparison?

8. Using the provided preference list, answer the questions that follow it.

	Number of Voters			
Rank	4	3	2	1
First Choice	D	B	A	A
Second Choice	B	A	C	C
Third Choice	A	C	D	B
Fourth Choice	C	D	B	D

a. How many voters were there?

b. How many voters have ranked the candidates in the order A, C, D, B?

c. Who is the winner using Plurality Voting?

d. Who is the winner using the Borda Count Method?

e. Who is the winner using the Hare System?

f. Who is the winner using Pairwise Comparison?

9. Using the provided preference list, answer the questions that follow it.

	Number of Voters			
Rank	4	2	2	1
First Choice	B	C	A	A
Second Choice	A	B	C	B
Third Choice	C	A	B	C

a. How many voters were there?

b. How many voters have ranked the candidates in the order B, A, C?

c. Who is the winner using Plurality Voting?

d. Who is the winner using the Borda Count Method?

e. Who is the winner using the Hare System?

f. Who is the winner using Pairwise Comparison?

10. Using the provided preference list, answer the questions that follow it.

	Number of Voters			
Rank	4	4	2	1
First Choice	A	C	B	A
Second Choice	B	B	C	C
Third Choice	C	A	A	B

a. How many voters were there?

b. How many voters have ranked the candidates in the order C, A, B?

c. Who is the winner using Plurality Voting?

d. Who is the winner using the Borda Count Method?

e. Who is the winner using the Hare System?

f. Who is the winner using Pairwise Comparison?

11. In a 3-candidate approval voting election with 14 voters, if 7 approve of A and B, 5 approve of B and C, and 2 approve of A and C, who wins the election?

12. In a 4-candidate approval voting election with 15 voters, if 2 approve of A and C, 7 approve of B and C, and 6 approve of A and D, who wins the election?

13. Use the provided table of votes and the Approval Voting Method to answer the questions that follow it.

	Specific Voters									
	#1	#2	#3	#4	#5	#6	#7	#8	#9	#10
Candidate A	✓	✓			✓		✓		✓	✓
Candidate B	✓	✓	✓			✓		✓		
Candidate C		✓			✓				✓	
Candidate D	✓		✓	✓	✓		✓	✓		✓
Candidate E		✓	✓			✓			✓	
Candidate F				✓	✓		✓	✓		

a. Which candidate is chosen if only one person is to be elected Chairman of the Board?

b. Which candidates will be elected if three people are being elected to the Board?

c. Which candidates are chosen if 60% approval is needed to be elected?

339

Answers to Section 6.2 Exercises

1. a. iv   b. i   c. ii   d. iii   e. v

2. 3

3. 15

4. Carol

5. a. 8   b. 3   c. D   d. B   e. C   f. B

6. a. 9   b. 1   c. B   d. B   e. A
   f. There is no winner. (3-way tie between A, B, and C)

7. a. 11   b. 5   c. C   d. B   e. C
   f. There is no winner. (3-way tie between B, C, and D)

8. a. 10   b. 2   c. D   d. A   e. D
   f. There is no winner. (2-way tie between A and B)

9. a. 9   b. 4   c. B   d. B   e. B   f. B

10. a. 11   b. 0   c. A   d. B   e. C   f. B

11. B

12. C

13. a. D   b. A, B, and D c. A and D

6.3 Dummies & Dictators (Weighted Voting Systems)

Weighted Voting

So far, in all the voting methods we have discussed, all voters were treated equally. However, not all voting systems operate that way. A common example is a situation where we have stockholders who are given voting power based on the number of shares of stock they own. A system in which the voters are not treated equally is a **weighted voting system**.

In a weighted voting system, each participant has a number of votes, which is called his or her weight. In order to determine whether the result of a vote is "pass" or "fail," a **quota** is established. A quota must be made up of more than half of the votes. A common method for setting the quota is to use a **simple majority quota**. When this method is used, the quota is the smallest whole number greater than half of the total weight of the voters.

If the sum of the weights of the voters who are in favor of a proposal is greater than or equal to the quota, then the proposal passes. When this is the case, the group of voters is called a **winning coalition**.

When the group of voters that support a proposal has a combined weight less than the quota, the proposal does not pass. This group of voters is called a **losing coalition**.

Notation for Weighted Voting Systems

When we describe a weighted voting system, the voting weight of each participant, as well as the quota, is specified using the following notation. The quota is represented by q, and the weights of the individual voters are represented by $w_1, w_2, w_3, \ldots, w_n$. Also, the individual weights are always listed in order from highest to lowest.

The system is written:

$$[q : w_1, w_2, w_3, \ldots, w_n]$$

EXAMPLE 1: For the voting system [7 : 5, 3, 2], identify the quota, the number of voters, and the weight of each voter.

The quota is seven, and there are a total of three voters. The weight of the first voter is five, the weight of the second voter is three, and the weight of the third voter is two.

A **dummy voter** is a voter whose vote does not matter. The system [12 : 9, 3, 2] has a dummy voter. In this situation, the only way that a proposal can pass is if the nine-weight voter and the three-weight voter support it, and in this case the support of the two-weight voter is not necessary. Since the support of the two-weight voter is not needed to pass a proposal, and the two-weight voter cannot prevent the other voters from passing a proposal, this voter has no impact on the outcome of a vote.

EXAMPLE 2: Given the weighted voting system [9 : 6, 3, 1], is there a dummy voter?

Yes. The one-weight voter is a dummy voter.

EXAMPLE 3: Given the weighted voting system [10 : 6, 3, 1], is there a dummy voter?

No. All of the voters actually have equal power, because in order for a proposal to pass, it must be supported by *all* of the voters. In other words, in the [10 : 6, 3, 1] system, all three voters form a coalition.

EXAMPLE 4: Given the weighted voting system [23 : 21, 20, 4], is there a dummy voter?

No. At first glance, we may think that the 4-weight voter would not have much power. However, since any two voters can reach the quota, the 4-weight voter actually has the *same* power as the other voters in this system.

A **dictator** is a voter who has enough power to determine the outcome of any vote, regardless of how the others cast their votes. The system [35 : 40, 10, 5] has a dictator, because the 40-weight voter has enough weight to pass a proposal, and the other voters cannot pass a proposal without the consent of the 40-weight voter. When there is a dictator, all the other voters are dummy voters.

A voter whose support is necessary to pass a proposal has **veto power**. The system [8 : 5, 3, 1] actually includes two voters with veto power. Since a proposal cannot reach this quota without the support of the five-weight voter and the three-weight voter, these voters have veto power.

Note the difference between a dictator and a voter with veto power. In order to be a dictator, the voter must have a weight greater than or equal to the quota. Since the quota must be greater than half of the total weight of the voters, there can never be more than one dictator, if one exists at all.

On the other hand, a voter with veto power can prevent a proposal from passing, but if this voter is too weak to be a dictator, he cannot pass a proposal without help.

EXAMPLE 5: Given the weighted voting system [71 : 40, 35, 31, 24, 3], answer the following.

133

How many voters are in the system?

a. What is the quota?

b. Is there a dictator?

c. Is there a dummy voter?

d. How many voters have veto power?

e. How many voting coalitions are there with exactly the quota required for the proposal to pass?

Answers

a. Five

b. 71

c. No.

d. Yes. The 3-weight voter.

e. None.

f. One. The only way to get a total weight of exactly 71 is if the 40-weight and 31-weight voters are the only ones who support the proposal.

EXAMPLE 6: Given the weighted voting system [30 : 27, 10, 4], answer the following.

    a. How many voters are in the system?

    a. What is the quota?

    b. Is there a dictator?

    c. Is there a dummy voter?

    d. How many voters have veto power?

    e. How many voting coalitions are there with exactly the quota required for the proposal to pass?

Answers

    a. Three

    b. 30

    c. No.

    d. No.

    e. One. The 27-weight voter is needed in any winning coalition.

    f. None.

Important Note

In the examples provided in this section, the systems only had a few voters. In most weighted voting systems, however, there are many, many voters. Typically, a company has thousands of voters, which are often weighted by the number of shares of company stock each voter owns. The same principals still exist, though; quotas must still consist of more than half of the weighted votes, and a dictator will have a weight greater than the quota. Even though it is not uncommon for single voters to be dummy voters, many coalitions are usually formed. Keep in mind, though, if a company has a total of a 50,000-weight voting system, a quota must consist of more than 25,000 votes. In such a case, we often see CEOs with weights in excess of 20,000 votes.

Another common practice is for non-voting members to transfer their voting authority to a different member. This happens with many homeowner's associations—members who do not vote may have their votes conceded to the Association President, which can turn the President into a dictator. If, indeed, this is an option, it will be spelled out in the by-laws of the association.

Section 6.3 Exercises

1. Determine the quota for a voting system that has the following number of votes and uses a simple majority quota.

    a. 26 = 13

    b. 66 = 33

    c. 83 = 42

    d. 44 = 22

2. For the weighted voting system [9 : 9, 5, 2], indicate whether the following statements are true or false.

    a. There are four voters in this system.

    b. The nine-weight voter is a dictator.

    c. The two-weight voter is a dummy.

3. For the weighted voting system [11 : 5, 5, 1, 1, 1], indicate whether the following statements are true or false.

    a. Three of these voters have equal weight.

    b. None of these voters have veto power.

    c. There is exactly one dummy voter.

4. For the weighted voting system [8 : 8, 2, 2, 1], indicate whether the following statements are true or false.

   a. There is a dictator.

   b. There are exactly three dummy voters.

   c. Exactly one voter has veto power.

5. Given the weighted voting system [61 : 35, 30, 25, 2]:

   a. How many voters are in the system?

   b. What is the quota?

   c. Is there a dictator?

   d. If there are any dummy voters, identify them.

   e. If there are any voters with veto power, identify them.

6. Given the weighted voting system [38 : 27, 10, 5, 3, 1]:
   a. How many voters are in the system?

   b. What is the quota?

c. Is there a dictator?

d. If there are any dummy voters, identify them.

e. If there are any voters with veto power, identify them.

7. Given the weighted voting system [6 : 5, 3, 2, 1]:

a. How many voters are in the system?

b. What is the quota?

c. Is there a dictator?

d. If there are any dummy voters, identify them.

e. If there are any voters with veto power, identify them.

8. Given the weighted voting system [4 : 4, 2, 1]:

a. How many voters are in the system?

b. What is the quota?

c. Is there a dictator?

d. If there are any dummy voters, identify them.

e. If there are any voters with veto power, identify them.

9. Given the weighted voting system [10 : 9, 2, 2, 2, 2]:

   a. How many voters are in the system?

   b. What is the quota?

   c. Is there a dictator?

   d. If there are any dummy voters, identify them.

   e. If there are any voters with veto power, identify them.

10. Given the weighted voting system [59 : 35, 35, 25, 2]:

    a. How many voters are in the system?

b. What is the quota?

c. Is there a dictator?

d. If there are any dummy voters, identify them.

e. If there are any voters with veto power, identify them.

11. Given the weighted voting system [17 : 7, 5, 3, 2]:

    a. How many voters are in the system?

    b. What is the quota?

    c. Is there a dictator?

    d. If there are any dummy voters, identify them.

    e. If there are any voters with veto power, identify them.

12. Given the weighted voting system [12 : 6, 4, 4, 2], how many voting coalitions are there with exactly the quota required to win?

13. Given the weighted voting system [10 : 6, 4, 4, 2], how many voting coalitions are there with exactly the quota required to win?

14. Given a system with four voters, having weights of 30, 29, 28, and 13, of which only one voter has veto power:

    a. Which voter has veto power?

    b. What is the quota?

    c. Is there a dummy voter?

15. If you hold 51% of the stock in a company, you have total control over the company. Why?

16. Why does having a dictator make all of the other voters dummies?

Answers to Section 6.3 Exercises

1. a. 14   b. 34   c. 42   d. 23

2. a. False   b. True   c. True

3. a. True   b. False   c. False

4. a. True   b. True   c. True

5. a. 4   b. 61   c. No.   d. None.
   e. The 35-weight voter has veto power.

6. a. 5   b. 38   c. No.   d. None.
   e. Both the 10- and the 27-weight voters have veto power.

7. a. 4   b. 6   c. No.   d. None.
   e. None.

8. a. 3   b. 4   c. Yes.   d. The two- and one-weight voters.
   e. The four-weight voter.

9. a. 5   b. 10   c. No.   d. None.
   e. The nine-weight voter.

10. a. 4   b. 59   c. No.   d. The two-weight voter.
    e. None.

11. a. 4   b. 17   c. No.   d. None.
    e. All the voters have veto power.

12. 2

13. 3

14. a. The 30-weight voter.

   b. 71. With veto power, the 30-weight voter must be included in a winning coalition. Thus, the quota must be greater than 29+28+13 = 70. And, if the quota was 72, both the 29- and 30-weight voters would have veto power.

   c. No.

15. Having 51% of the stock makes you equivalent to a 51-weight voter in a system with a quota of 51. This makes you a dictator.

16. The dictator can pass or reject a motion alone, which makes the votes of all of the other voters irrelevant.

6.4 Who Gets the Bigger Half? (Fair Division)

Fair division is the concept of dividing something between two or more people in such a way that each person finds his/her share to be fair. There are a number of ways to achieve fair division, and some of them will be investigated in this section. Before we begin, however, we must clarify one very important point. Divisions do not have to be equal; they just need to be fair in the eye of the beholder. Each party should get what he or she deems to be a fair share, and what happens to everyone else does not matter. Also, the parties involved in the division are expected to be reasonable. Otherwise, an independent arbitrator must oversee the procedure.

> Fair division *does not* mean equal division.

Discrete items, such as cash or a box full of CDs, can usually be split equally. Some **continuous items**, such as pets and cars cannot be cut up, as they would lose their value and become worthless. Other continuous items, like pizza, can be split up without losing value. Many times, continuous items that cannot be split are liquidated into discrete items, such as money, to complete a division.

Fair Division Between Two People

Divider-Chooser Method

This is a practical (and simple) approach to having two people share an object that can be cut into pieces. You may already be familiar with this process.

Let's say two children, a brother and sister, want to share a piece of cake. While this could result in a terrible argument, there is a simple, yet fair way to divide that piece of cake.

1. First, flip a coin to determine which child will be Player 1. The other will be Player 2.

2. Next, Player 1 gets to cut the piece of cake into two parts, wherever he/she wishes.

3. Then, Player 2 gets to choose the piece of cake he/she would like to have, leaving the other piece for Player 1.

In this system, one player gets to divide the cake and the other player gets to choose which piece to take.

Taking Turns

One of the most basic approaches to the fair division of items is **taking turns**. While taking turns has inherent flaws, such as determining who will go first, when both parties know the preferences of the other, there is an interesting strategy that can be employed to make it a little fairer. Developed by mathematicians D.A. Kohler and R. Chandrasekaran in 1969, the process is known as the bottom-up strategy, and it is an excellent way to divide items when a neutral arbitrator is used.

Bottom-Up Strategy

When two parties are going to divide items, simply taking turns can often leave an individual with an undesirable item. When each party knows the preferences of the other, mapping out the picks from the bottom-up can help both parties achieve satisfaction, and this process is known as the **bottom-up strategy**. When this strategy is used, the individual identified with the last choice begins by each putting the item from the *bottom* of the other person's list as their *bottom choice*. This process continues, with the parties taking turns in reverse order, until all of the choices have been mapped out.

The most confusing part about the bottom-up strategy can be which pick to assign first. When we say, "Bob gets to pick first," we mean, if Bob and Tom were to divide the items normally, Bob would select one of the items to be his, and they would continue taking turns until all the items were gone. So, if there were four items, Bob would have the first pick, Tom would get the second pick, Bob would get the third overall pick, and Tom would take what was left with the fourth and final pick. To get a better grasp of this concept, we can literally map out these picks, providing a blank for each item to be selected.

1. Bob: _____

2. Tom: _____

3. Bob: _____

4. Tom: _____

Notice how we can make that "map" without actually knowing what items they are dividing; we just know there are four of them (and, hence, four picks).

Next, in order to use the bottom-up strategy, we need to know what the items are and how Bob and Tom rank them. To continue with our example, let's say they provide the following rankings for a dog, iPod, couch and a set of books.

Tom	Bob
Dog	iPod
iPod	Couch
Couch	Dog
Books	Books

Then we literally assign the picks by filling in our blanks in the map—from the bottom-up. That is, since Tom would be picking last, according to the bottom-up strategy, he places the lowest ranked item from *Bob's list* in that last spot. According to the lists, that pick will be the books.

1. Bob: _____

2. Tom: _____

3. Bob: _____

4. Tom: Books _____

To avoid accidentally selecting them again, with the books being assigned to Tom, cross "Books" off *both* lists.

Tom	Bob
Dog	iPod
iPod	Couch
Couch	Dog
~~Books~~	~~Books~~

Continuing from the bottom-up, the next step is to assign the next-to-last overall pick, which, in this case, is the #3 pick. Since this pick belongs to Bob, he takes the lowest available item from Tom's list, which is the couch. So, put the couch in the blank for the #3 pick and cross it off the lists.

1. Bob: _____

2. Tom: _____

3. Bob: <u>Couch</u>

4. Tom: <u>Books</u>

Tom	Bob
Dog	iPod
iPod	~~Couch~~
~~Couch~~	Dog
~~Books~~	~~Books~~

Continuing this process, we will see Tom gets the dog (Bob's lowest rated available item), and then, Bob will select the only thing left, which is the iPod. In the end, Bob will get the iPod and the couch, while Tom takes home the dog and the books. By agreeing to use the bottom-up strategy, Tom ends up with his first and fourth ranked items (even though he picked second), which is a good result for him, and Bob gets his first and second choices, which is a great result for him.

EXAMPLE 1: Ben and Jerry will use the bottom-up strategy to divide an iPod, speakers, a DVD player, and a turntable. Ben gets to choose first. Each person ranked the items, in order of preference before the selection process begins.

How will the items be divided?

Ben	Jerry
iPod	DVD Player
Speakers	iPod
Turntable	Speakers
DVD Player	Turntable

Since Ben will choose first, the picks will go as follows:

1. Ben: _____

2. Jerry: _____

3. Ben: _____

4. Jerry: _____

Using the bottom-up strategy, we start by identifying what should be the last item chosen. Thus, with the fourth (and final) pick, Jerry indicates he will choose the item Ben wants the least. In other words, Jerry will choose the item from the *bottom* of Ben's list.

1. Ben: _____

2. Jerry: _____

3. Ben: _____

4. Jerry: <u>DVD Player</u>

Next, as the third pick, Ben will slot in the *lowest available item* from Jerry's list.

1. Ben: _____

2. Jerry: _____

3. Ben: <u>Turntable</u>

4. Jerry: <u>DVD Player</u>

Moving up the list (going "bottom-up"), Jerry then indicates the lowest available item on Ben's list. Since the DVD Player and the turntable have already been chosen, the next available item on Ben's list is the speakers.

1. Ben: _____

2. Jerry: Speakers_____

3. Ben: Turntable_____

4. Jerry: DVD Player_____

Finally, Ben will slot in his first pick, which will be the lowest available item on Jerry's list.

1. Ben: iPod_____

2. Jerry: Speakers_____

3. Ben: Turntable_____

4. Jerry: DVD Player_____

Using the bottom-up strategy, Ben will end up with the iPod and the turntable. Jerry will end up with the speakers and the DVD Player.

In the previous example, both parties got their first and third choices, which is a more fair division than simply picking items 1-2-3-4. If they had simply taken turns from the beginning, Ben would have gotten his top two choices, but Jerry would have ended up with his top choice and his least desirable item.

It is also important to note, however, that the party allowed to choose first definitely makes a difference in how the items will be divided. Let's look at the previous example again, except this time let's see what happens when Jerry goes first.

EXAMPLE 2: Assume Ben and Jerry use the bottom-up strategy to divide an iPod, speakers, a DVD player, and a turntable. Jerry gets to choose first. Each person ranked the items, in order of preference, before the selection process begins.

How will the items be divided?

Ben	Jerry
iPod	DVD Player
Speakers	iPod
Turntable	Speakers
DVD Player	Turntable

Jumping to the completed map of picks, we have:

1. Jerry: <u>iPod</u>

2. Ben: <u>Speakers</u>

3. Jerry: <u>DVD Player</u>

4. Ben: <u>Turntable</u>

This time, Ben will end up with the speakers and the turntable, while Jerry will get the iPod and the DVD Player.

With division in Example 2, Jerry was able to get his first and second choices, while Ben got his second and third choices. Since neither person was stuck receiving the fourth item on his list, this is a division both parties can find to be agreeable.

The Adjusted Winner Procedure

The **Adjusted Winner Procedure** is a method by which two parties can divide a group of items in a fair manner, provided one of the items can be divided.

This process uses the following steps:

1. Each party is allowed to assign a total of 100 points to the items being divided. The more important an item is to a party, the more points it should be given.

2. Initially, whichever party assigns the highest point value to a specific item is given that item. (We will assume there are no ties in the point assignments.)

3. At this point, the number of points each party has received is determined. If there is any item assigned an equal number of points by the two parties, the party who has currently received the lower number of points is given that item.

4. Once the initial division is done, if the point values are equal, the process is complete. However, this is very unlikely. It is probable that one party has received more points than the other, and this will have to be rectified.

5. Let's call the party that has received more points Party A and the other Party B. In order for this division to be fair, we will need to take some points away from Party A and give them to Party B. To do so, we may have to transfer a fractional amount of an item from Party A to Party B. The item to be split either needs to be split without losing its value or needs to be liquidated to perform the division.

Disclaimers for the Adjusted Winner Procedure

In the second step, as indicated in the parentheses, we are keeping the process simpler by assuming there are no ties in the point assignments. In the event of a tie that cannot be resolved by the parties involved, then an independent arbitrator can be used.

In the last step, it is necessary for us to point out there are different methods that can be used to determine which item gets split up in order to equalize the points. One method allows for the point leader to determine which item gets divided, and, typically, he will split the least valuable item in his list. If, however, there is an item that is easier to divide (such as cash), the division could be done with that.

EXAMPLE 3: Champ and Monte have purchased the contents of an abandoned storage shed. They look inside and find some sporting goods, stuffed animals, and textbooks. They decide to create a fair division of these objects using the Adjusted Winner Procedure, and they assign their points as shown:

	Point Allocations	
Item	Champ	Monte
Sporting Goods	60	10
Stuffed Animals	10	70
Textbooks	30	20

Initially, because the parties assigned higher point values to these items, Champ is given the textbooks and the sporting goods, while Monte is given the stuffed animals.

At this point, Champ has been given 90 points worth of the items, and Monte has been given items that he deemed to be worth 70 points.

Given that the textbooks are the least valuable item to Champ, he will balance this division by keeping part of them and giving the rest to Monte. That is, each party will get a *fraction* of the books. To calculate the appropriate fractions, let's start by saying Champ will get "x." Monte will get the fraction of the books that are not given to Champ. This amount will be represented by "1 – x."

We will multiply the points Champ assigned to the books by x to find out the number of points he will get from the books. Similarly, we will multiply the number of points Monte assigned to the books by 1 – x to find the number of points he will receive from the books.

Using the point values Champ and Monte assigned to the items, we can set up an expression that shows the total number of points each party will have after dividing the books.

	Points Before Books	Points From Books	Total Points Received
Champ	60	30x	**60 + 30x**
Monte	70	20(1 – x)	**70 + 20(1 – x)**

To achieve fairness in this division, the total points received by each of Champ and Monte need to be equal. Thus, we set those two expressions equal to one another and solve for x. That will give us the fraction of the textbooks to be given to Champ.

$$60 + 30x = 70 + 20(1 – x)$$

$$60 + 30x = 70 + 20 – 20x$$

$$60 + 30x = 90 – 20x$$

$$50x = 30$$

$$x = 3/5$$

So, Champ will be given 3/5 of the textbooks, and the remaining 2/5 of the books will be given to Monte.

To illustrate *how* this achieves fair division, let's take a look at the total number of points given to each party. Champ was given the sporting goods, worth 60 points. He was also given 3/5 of the textbooks, which, to him, is worth $(3/5)(30) = 18$ points. So, all together, Champ was given items that added to 78 points.

Monte was given the stuffed animals, worth 70 points. He was also given 2/5 of the textbooks, which, to him, is worth $(2/5)(20) = 8$ points. So, all together, Monte was given items that added to 78 points.

Since each party received the same number of points, this is a fair division of the items.

EXAMPLE 4: Dave and Dan have to divide four items: a riding lawn mower, a pool table, a couch, and a CD collection. Using the Adjusted Winner procedure, Dave assigns his allotment of 100 points as follows: 40 on the mower, 10 on the pool table, 20 on the couch, and 30 on the CDs. Dan assigns his points as follows: 20 on the mower, 50 on the pool table, 10 on the couch, and 20 on the CDs. Indicate how the items should be divided.

	Point Allocations	
Item	Dave	Dan
Lawn Mower	40	20
Pool Table	10	50
Couch	20	10
CDs	30	20

Dan would get the pool table (50 points) and Dave would take the lawn mower (40) and the couch (20). Since allowing either party to take the entire CD collection would result in an unfair division, the CD collection is the simplest item to split up, as it is the only item that can be "divided" without being liquidated.

Letting x be the fraction of the CDs Dave will get, we can use the point assignments to compute the amount.

	Points Before CDs	Points from CDs	Total Points Received
Dave	60	$30x$	$60 + 30x$
Dan	50	$20(1-x)$	$50 + 20(1-x)$

Dave has 60 points from the mower and couch, and will get 30x points from the CDs. Dan has 50 points from the pool table, and will get 20(1-x) points from the CDs. Setting these point totals equal to each other and solving for x, we have:

$$60 + 30x = 50 + 20(1 - x)$$

$$60 + 30x = 50 + 20 - 20x$$

$$50x = 10$$

$$x = 1/5$$

Dave gets 1/5 of the CDs and Dan gets the remaining 4/5 of them.

Checking the point totals,

Dave: 40 (from the lawn mower) + 20 (couch) + 30 × 1/5 (CDs) = 66

Dan: 50 (pool table) + 20 × 4/5 = 50 + 16 = 66

The point totals are equal, so the division is fair.

Fair Division Between Two or More People

Knaster Inheritance Procedure

When more than two people are involved in a division, the previous methods can become quite complicated. An alternative method, developed by Polish mathematician Bronislaw Knaster in 1945, the **Knaster Inheritance Procedure** (also called **the method of sealed bids**) is completed using the following principles:

1. Each person writes down the highest amount they would be willing to pay for each item—without knowing what the other parties bid for the same item. Assume there are no ties in the bids.

2. The highest bidder for each item will win it, and the other people will be compensated with money, paid to them by the person who won the item. Bidders, however, are not compensated directly. Instead, if there are n bidders, the winner will place $(n-1)/n$ × (the

winning bid) into a kitty. For example, if there are 4 bidders, the winner places 3/4 of his bid into the kitty.

3. The non-winners will then take 1/nth of their individual bids from the kitty. For example, if three people are involved in the division, each person who did not win is entitled to 1/3 of the amount he/she bid for that item.

4. After each non-winner has taken the appropriate amount from the kitty, the remaining funds in the kitty are then divided equally among each bidder, including the one who won the item.

EXAMPLE 4: There are two items, a painting and a sculpture, to be divided fairly among three people using the method of sealed bids.

Each person puts forth a bid on each item, as follows:

Item	Audrey	Brian	Cindy
Painting	4,200	6,000	5,100
Sculpture	6,600	4,800	7,500

When using the Knaster Inheritance Procedure, each item is handled separately.

Starting with the painting …

Brian is the high bidder, so he will get the painting.

There are three people involved in this division, so Brian must place 2/3 of the $6,000 bid into a kitty. $2/3 \times \$6,000 = \$4,000$

Audrey and Cindy, since they did not get the painting, will take money from the kitty. Their fair shares of the painting are equal to 1/3 of the amount of their respective bids.

Audrey bid $4,200 on the painting, so she is entitled to 1/3 of $4,200, which is $1,400. Thus, she takes $1,400 from the kitty.

Cindy bid $5,100 on the painting, so she is entitled to 1/3 of $51,00, which is $1,700. So, she takes $1,700 from the kitty.

The kitty started with $4,000. After Audrey and Cindy each removed their shares, $900 remained in the kitty. That $900 is divided equally among all three people, so each of them receives $300.

At this point, the Knaster Procedure has taken care of division of the painting:

- Audrey's share: $1,400 + $300 = $1,700

- Brian's share: painting − $4,000 + $300 = painting − $3,700

- Cindy's share: $1,700 + $300 = $2,000

The same procedure will be used with the sculpture.

Cindy is the high bidder, so she will get the sculpture.

There are three people involved in this division, so Cindy must place 2/3 of the $7,500 bid into a kitty. $2/3 \times \$7,500 = \$5,000$

Audrey and Brian, since they did not get the sculpture, will take money from the kitty. Their fair shares of the painting are equal to 1/3 of the amount of their respective bids.

Audrey bid $6,600 on the sculpture, so she is entitled to 1/3 of $6,600, which is $2,200. Thus, she takes $2,200 from the kitty.

Brian bid $4,800 on the sculpture, so she is entitled to 1/3 of $4,800, which is $1,600. So, he takes $1,600 from the kitty.

This leaves the kitty with $5,000 − $2,200 − $1,600 = $1,200, which is divided equally among the three people. So, each of them receives $400.

With regard to the sculpture:

- Audrey's share: $2,200 + $400 = $2,600

- Brian's share: $1,600 + $400 = $2,000

- Cindy's share: sculpture − $5,000 + $400 = sculpture − $4,600

Finally, we combine the results of dividing the two items to obtain the final distribution.

	Painting	Sculpture	Total
Audrey	$1,700	$2,600	**$4,300**
Brian	Painting—$3,700	$2,000	**Painting—$1,700**
Cindy	$2,000	Sculpture—$4,600	**Sculpture—$2,600**

So, to complete the fair division of these items at the same time, Brian can pay $1,700 to Audrey and gets the painting, while Cindy pays $2,600 to Audrey and gets the sculpture.

EXAMPLE 5: Jenny and Amy are no longer going to be roommates, but they have to decide who will get to keep their dog. Obviously, they won't be able to "divide" the dog, and they do not want to sell it, either. So, they will use the Knaster Procedure instead.

- Jenny "bids" $300 for the dog.

- Amy "bids" $400 for the dog.

Since this division involves two people, each of them is entitled to 1/2 of the value of their bid.

Amy, as the higher bidder, will get the dog, and she will have to put 1/2 of her bid, $200, into the kitty.

Jenny takes 1/2 of her bid, $150, from the kitty.

This leaves $50 in the kitty, which is split evenly between the two people. So each woman will take $25 more from the kitty.

That means, to be fair, Amy gets the dog and pays $175 to Jenny.

In the Knaster Procedure, since the bids in the procedure are monetary, there is no need to worry about whether or not the items can be divided.

Section 6.4 Exercises

1. Two friends want to share the last cookie. Describe how it may be divided using the divider-chooser method.

2. Butch and Droopy must make a fair division of these objects using the Adjusted Winner Procedure, and they assign points as shown.

	Point Allocations	
Item	Butch	Droopy
Dynamite	50	10
Sports Car	30	60
Gold Coins	20	30

Butch will end up with the dynamite and a fraction of the gold coins. What fraction of the gold coins will he get?

3. Labor and management are negotiating the following items. Using the Adjusted Winner Procedure, they assign points as shown.

Point Allocations

Issue	Labor	Management
Benefits	45	50
Salary	35	40
Vacation	20	10

Management will be awarded all the benefits issues and a percentage of the salary issues. What percentage of the salary issues will they get?

4. Jack and Diane must make a fair division of these possessions using the Adjusted Winner Procedure, and they assign points as shown.

Point Allocations

Item	Jack	Diane
Horse Farm	49	16
Urban Condo	16	63
House Boat	35	21

Jack will end up with the Horse Farm and a percentage of the value of the House Boat. What percentage of the House Boat will he get?

5. Three brothers will use the Knaster inheritance procedure to divide the following items. What will the final distribution be?

Item	David	Ethan	Frank
Car	11,400	13,500	10,200
Boat	9,300	7,500	8,400

6. The semester has ended, and three roommates need to divide two items—old couch and a fairly new television. They will use the method of sealed bids, and each person puts forth a bid on each item, as follows.

Item	Mickey	Willie	Ted
Couch	90	120	78
TV	330	414	480

What will be the final distribution of these items?

7. A singing quartet has broken up. The four members will use the Method of Sealed Bids to divide their private plane and tour bus, with the bids as indicated below. What will the final distribution be?

Item	Soprano	Alto	Tenor	Bass
Plane	76,000	80,000	68,000	72,000
Bus	52,000	48,000	64,000	68,000

8. In the process of a divorce, Jeff and Charlotte are going to use the Method of Sealed Bids to determine who will get their car. Jeff bids $24,000 and Charlotte bids $21,500, so Jeff will get the car. How much will he have to pay Charlotte in order to keep the division fair?

9.   Assume Brian and Daryl use the bottom-up strategy to divide the different types of sports cards listed below, and Brian gets to choose first. What items will each person receive?

Brian	Daryl
Hockey	Football
Football	Baseball
Baseball	Basketball
Basketball	Hockey

10. Assume Brian and Daryl use the bottom-up strategy to divide the different types of sports cards listed below, and Daryl gets to choose first. What items will each person receive?

Brian	Daryl
Hockey	Football
Football	Baseball
Baseball	Basketball
Basketball	Hockey

11. Assume Alvin and Simon use the bottom-up strategy to divide the items listed below, and Alvin gets to choose first. What items will each person receive?

Alvin	Simon
Boat	Car
Truck	Truck
Car	Boat
Computer	TV
Furniture	Computer
TV	Furniture

12. Assume Alvin and Simon use the bottom-up strategy to divide the items listed below, and Simon gets to choose first. What items will each person receive?

Alvin	Simon
Boat	Car
Truck	Truck
Car	Boat
Computer	TV
Furniture	Computer
TV	Furniture

Answers to Section 6.4 Exercises

1. They should flip a coin to determine the person who will cut the cookie. Then, the other person can select either of the two pieces.

2. 4/5

3. Management will get 1/15 of the salary demands. In other words, Labor will get 14/15 (about 93%) of the salary they desire.

4. 62.5%

5. David will get the boat and pay $1,500.
    Ethan will get the car and pay $5,600.
    Frank will receive $7,100.

6. Mickey will get $172.
    Willie will get the couch and receive $90.
    Ted will get the television and pay $262.

7. Soprano will get $36,000.
    Alto will get the plane and pay $44,000.
    Tenor will get $37,000.
    Bass will get the bus and pay $29,000.

8. $11,375

9. Brian will get the football and hockey cards.
    Daryl will get the baseball and basketball cards.

10. Daryl will get the football and basketball cards.
    Brian will get the hockey and baseball cards.

11. Alvin will end up with the truck, the boat, and the furniture.
    Simon will end up with the car, the TV, and the computer.

12. Simon will get the truck, the car, and the TV.
    Alvin will get the boat, the computer, and the furniture.

6.5 Carving Up the Turkey (Apportionment)

Apportionment

Apportionment is the process of dividing something into shares that are proportional to the characteristics of the situation. An example of this is the House of Representatives of the United States. In the House, states with higher populations are given more representatives than states with lower populations.

Solving an apportionment problem should involve an unbiased and repeatable process. A biased process would disproportionately favor one political party or state, and an example of a non-repeatable process would be pulling the results from a hat (that is, it would be unlikely to produce the same results in consecutive attempts.) While there are a few different unbiased and repeatable processes that can be used, we will focus our study on a method proposed by the first Secretary of the Treasury of the United States, Alexander Hamilton.

EXAMPLE 1: The math department at a local college has hired two new faculty members who, together, will teach a total of 10 sections of algebra, calculus, and liberal arts math. The department chair is not sure how many sections of each course to assign to these new teachers, so she has gathered some data about student enrollment. There are currently 220 students enrolled in these courses in the following manner.

Course	Enrollment
Algebra	103
Calculus	34
Liberal Arts Math	83

Although many apportionment problems do not involve the House of Representatives, it is common to use terminology from that setting. The following are some terms used when solving apportionment problems, set in the context of this problem:

- **Population**: The total number of students enrolled in these courses.

- **House Size**: The total number of sections.

- **Standard Divisor**: The mean number of students per section.

- **Quota**: The exact number of sections of each course that would be offered, if a whole number of sections were not required. This is also referred to as the **calculated quota**.

To determine the number of sections of each course that should be offered, the first thing we do is calculate the standard divisor. The standard divisor is found using the following formula.

Standard Divisor = (population)/(house size)

Since the population (number of students) is 220, and the house size (number of sections) is 10, the standard divisor for this example will be $220/10 = 22$.

To find the quota for each class, divide the corresponding population by the standard divisor. Using the numbers from our enrollment data, we see the following.

Course	Quota
Algebra	$103/22 = 4.68$
Calculus	$34/22 = 1.55$
Liberal Arts Math	$83/22 = 3.77$

Since the department chair cannot offer fractions of sections, these quotas will have to be modified to determine the number of sections to offer.

Rounding might seem like a reasonable method for determining the number of sections. However, if we round each quota to the nearest whole number, the department will offer five sections of algebra, two sections of calculus, and four sections of liberal arts math, and the two instructors cannot cover a total of 11 sections.

For one possible solution, the department chair might decide a calculus class of 34 is too many and make certain there are two sections of calculus. Then, for the other eight sections, because there are more students enrolled in algebra, she may create five sections of algebra and three sections of liberal arts math.

This may be a reasonable approach, but another person may approach the situation differently. A standard, repeatable process is needed to fairly determine the offered sections.

The Hamilton Method

Alexander Hamilton devised this method of dealing with fractional quotas, which he called the "method of largest fractions." Using the language associated with the House of Representatives, this process involves three steps:

1. Determine the **calculated quota** for each state by dividing the state's population by the standard divisor.

2. Drop the fractional part of the quota, and assign each state the number of seats shown by the whole number. This is referred to as the **lower quota**.

3. If this does not fill all of the seats, then the remaining seats, called **surplus seats**, are given out, one at a time, to the states with the largest fractional parts.

Let's apply the Hamilton Method to the Example 1, by finding the calculated quota for each class and then dropping the fractional part (round it down) for the lower quota.

Course	Quota	Lower Quota
Algebra	$103/22 = 4.68$	4
Calculus	$34/22 = 1.55$	1
Liberal Arts Math	$83/22 = 3.77$	3

To begin with, there will be four sections of algebra, one section of calculus, and three sections of liberal arts math. This accounts for 8 of the 10 sections. In order to assign the other two sections, we look at the fractional parts of the quotas.

Liberal arts math has the highest fractional part, 0.77, and is given one of the two remaining sections. The next highest fractional part is associated with algebra, at 0.68, so it will get the final remaining section.

In this situation, using the Hamilton Method, the final apportionment will be:

• There will be five sections of Algebra.

• There will be one sections of Calculus.

• There will be four sections of Liberal Arts Math.

When the Hamilton Method is used, every apportionment is either equal to the lower quota or the whole number immediately greater than it, which is called the **upper quota**. A process that follows this condition for apportionment is said to satisfy the **quota rule**.

EXAMPLE 2: A small country with three states has 40 seats in its Congress. The populations of the states are indicated below. Using the Hamilton Method, how many seats in the Congress will be apportioned to each state?

State	Population
Eastland	2,400
Centralia	1,950
Westerley	1,475

First, we need to find the standard divisor. The total population for the country is 5,825. Using the formula for the Standard Divisor, we find:

$$\text{Standard Divisor} = 5{,}825/40 = 145.625$$

Next, find the quota for each state and round it down.

State	Quota	Lower Quota
Eastland	2,400/145.625 = 16.48	16
Centralia	1,950/145.625 = 13.39	13
Westerley	1,475/145.625 = 10.13	10

We start by apportioning 16 seats to Eastland, 13 to Centralia, and 10 to Westerley. This accounts for 39 of the 40 seats in Congress, so there is still one seat to be apportioned. It will be given to the state with the quota that has the largest fractional part, which is Eastland.

The final apportionment will be:

- Eastland gets 17 seats.

- Centralia gets 13 seats.

- Westerley gets 10 seats.

Potential Problems with Hamilton's Apportionment Method

Apportionment procedures, although logical, may unintentionally create problems. One potential problem with Hamilton's method is that it does not guarantee each state will be given representation (for the U.S. House of Representatives, Article 1, Section 2, of the U.S. Constitution does that). It is possible for a state to have a population below the standard divisor and, hence, have a quota less than one. And, if the fractional portion of the state's quota is not large enough, the state may not even get one of the extra seats remaining after the initial apportionment. If Hamilton's method were still being used today, there are four U.S. States with populations lower than what would be the standard divisor—Alaska, North Dakota, Vermont, and Wyoming. Each of those states, however, would have a fractional portion large enough to get one of the extra seats.

Rather than run through examples of the U.S. House, which would deal with 50 states and 435 seats, we will use smaller examples to demonstrate some of these problems.

EXAMPLE 3: Use the Hamilton Method to determine the apportionment of 10 sections of algebra, calculus, and liberal arts math, based on the estimated demand for each course as follows:

Course	Enrollment
Algebra	124
Calculus	13
Liberal Arts Math	83

The total number of students is 220, and there are to be 10 sections. Thus, the standard divisor is $220/10 = 22$.

Finding the quota for each class and, then, dropping the fractional part, we have:

Course	Quota	Lower Quota
Algebra	124/22 = 5.64	5
Calculus	13/22 = 0.59	0
Liberal Arts Math	83/22 = 3.77	3

To begin with, there will be five sections of algebra and three sections of liberal arts math, but no section of calculus. And, since algebra and liberal arts math have the two highest fractional parts, those courses will each get one of the two remaining sections, and no section of calculus will be offered.

The final apportionment will be:

• Six sections of algebra

• No sections of calculus

• Four sections of liberal arts math

Simply put, there is more demand for extra sections of algebra and liberal arts math than there is for one section of calculus.

The Alabama Paradox

A paradox is a situation that seems logistically impossible but can still happen. Based on the results of the 1880 Census, it was discovered that adding a seat to the House of Representatives could actually cause a state (Alabama) to *lose* a seat. Logically, it seems as if the extra seat would go to the state with the next highest fractional part. In reality, however, the extra seat changes the standard divisor and, hence, the fractional portions of the quotas will also change. Because this involved the State of Alabama, it is called the **Alabama Paradox.**

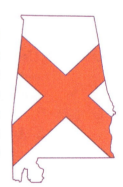

EXAMPLE 4: Use the Hamilton Method to apportion 29 seats to four states with the following populations.

State	Population
Eastland	150
Centralia	122
Mountania	38
Westerley	109

First, find the standard divisor.

$$\text{Total Population/Number of Seats} = 419/29 = 14.448$$

Next, we find the Quota for each state and round it down.

State	Quota	Lower Quota
Eastland	$150/14.448 = 10.38$	10
Centralia	$122/14.448 = 8.44$	8
Mountania	$38/14.448 = 2.63$	2
Westerley	$109/14.448 = 7.54$	7

Start by apportioning 10 seats to Eastland, 8 to Centralia, 2 to Mountania, and 7 to Westerley. This accounts for 27 of the 29 seats, so there are still two seats to be apportioned. They will be given to the two states with the largest fractional parts, which are Mountania (0.63) and Westerley (0.54).

The final apportionment will be as follows:

• Eastland gets 10 seats.

• Centralia gets 8 seats.

• Mountania gets 3 seats.

• Westerley gets 8 seats.

Now, leaving the populations the same, let's add another seat to the House.

EXAMPLE 5: Use the Hamilton Method to apportion 30 seats to 4 states with the following populations.

State	Population
Eastland	150
Centralia	122
Mountania	38
Westerley	109

Since there are 30 seats, the standard divisor is now:

Total Population/Number of Seats = 419/30 = 13.967

Next, we find the quota for each state and round it down.

State	Quota	Lower Quota
Eastland	150/13.967 = 10.74	10
Centralia	122/ 13.967 = 8.74	8
Mountania	38/13.967 = 2.72	2
Westerley	109/13.967 = 7.80	7

Start by apportioning 10 seats to Eastland, 8 to Centralia, 2 to Mountania, and 7 to Westerley. This accounts for 27 of the 30 seats, so there are still three seats to be apportioned. The highest fractional parts belong to Westerley (0.80), Eastland (0.74), and Centralia (0.74), so those states will get the three additional seats.

This final apportionment will be:

• Eastland gets 11 seats.

• Centralia gets 9 seats.

• Mountania gets 2 seats.

• Westerley gets 8 seats.

Comparing Examples 4 and 5, by adding a seat to the House and without changing the populations, Mountania has actually *lost* a seat. The **Alabama Paradox** occurs, because increasing the number of seats increases the fair share for larger states faster than it does for smaller states.

The Population Paradox

Another curious case that can occur with apportionment is the possibility of a state with an increasing population actually losing a seat to a state with a decreasing population. This is referred to as the **Population Paradox**.

EXAMPLE 6: Use the Hamilton Method to apportion 100 seats for the following 4 states.

State	Population
Eastland	34,300
Centralia	20,050
Mountania	1,250
Westerley	19,400

The standard divisor will be $75,000/100 = 750$.

Next, we find the quota for each state and round it down.

State	Quota	Lower Quota
Eastland	$34,300/750 = 45.73$	45
Centralia	$20,050/750 = 26.73$	26
Mountania	$1,250/750 = 1.67$	1
Westerley	$19,400/750 = 25.87$	25

We start by apportioning 45 seats to Eastland, 26 to Centralia, 1 to Mountania, and 25 to Westerley. This accounts for 97 of the 100 seats, so there are still three seats to be apportioned. They will be given to the states with the largest fractional parts, which are Westerley (0.87), Eastland (0.73), and Centralia (0.73).

The final apportionment will be:

• Eastland gets 46 seats.

• Centralia gets 27 seats.

• Mountania gets 1 seat.

• Westerley gets 26 seats.

Over time, the populations of these states change.

EXAMPLE 7: Use the Hamilton Method to apportion 100 seats for the following 4 states.

State	Population
Eastland	36,500
Centralia	20,100
Mountania	1,245
Westerley	19,420

We can see that the populations of Eastland, Centralia, and Westerley have increased, while the population of Mountania has decreased.

The new standard divisor is $77,265/100 = 772.65$

Next, we find the quota for each state and round it down.

State	Quota	Lower Quota
Eastland	$36,500/772.65 = 47.24$	47
Centralia	$20,100/772.65 = 26.01$	26
Mountania	$1,245/772.65 = 1.61$	1
Westerley	$19,420/772.65 = 25.13$	25

We start by apportioning 47 seats to Eastland, 26 to Centralia, 1 to Mountania, and 25 to Westerley. This accounts for 99 of the 100 seats, so there is still one seat to be apportioned. Since Mountania has the highest fractional part (0.61), it will be given the remaining seat.

The final apportionment will be as follows:

• Eastland gets 47 seats.

• Centralia gets 26 seats.

• Mountania gets 2 seats.

• Westerley gets 25 seats.

Comparing Examples 6 and 7, even though the populations of Centralia and Westerley are growing, they have each *lost* a seat. The population of Mountania is actually decreasing, but it has *gained* a seat. Because the total population changed, the standard divisor changed, and the fractional parts of the quotas changed. Thus, the extra seats got distributed differently.

The New State Paradox

A third paradox, the **New State Paradox**, is a bit more complicated. It can occur when new states are added to the House, and a proportional number of seats are added with them. Theoretically, the new states would likely get all of the new seats, and the other apportioned seats would remain the same. However, due to the rounding methods used, it is possible for an existing state (other than a new one) to *gain* a seat.

EXAMPLE 8: A homeowner's association (HOA) consists of two neighborhoods: the 296-resident Maplewood Acres and the 105-resident Oakey Oaks. Using Hamilton's Method, how are the 40 seats on the HOA Board apportioned?

There are a total of 401 residents for 40 seats. Thus, the standard divisor is 10.025. The corresponding quota for each state is:

Neighborhood	Quota	Lower Quota
Maplewood	296/10.025 = 29.53	29
Oakey Oaks	105/10.025 = 10.47	10

With the largest fractional portion of the quotas, Maplewood Acres would get the one extra seat.

• Maplewood Acres has 30 seats

• Oakey Oaks has 10 seats

Now look what can happen when a couple new neighborhoods are added to the HOA.

EXAMPLE 9: The HOA from Example 8 invites two new neighborhoods to join: The 52-resident Twin Pines and the 50-resident Ashville. With 102 new residents, the HOA adds 10 more seats to its board. Using the Hamilton Method, what is the new apportionment of the HOA?

There are now a total of 503 residents for 50 seats. Thus, the standard divisor is 10.06. The corresponding quota for each state is:

Neighborhood	Quota	Lower Quota
Maplewood	296/10.06 = 29.42	29
Oakey Oaks	105/10.06 = 10.44	10
Twin Pines	52/10.06 = 5.17	5
Ashville	50/10.06 = 4.97	4

With the largest fractional portions of the quotas, Ashville gets one seat and Oakey Oaks gets the other.

- Maplewood Acres has 29 seats

- Oakey Oaks has 11 seats

- Twin Pines gets 5 seats

- Ashville gets 5 seats

Comparing Examples 8 and 9, we see how the addition of two neighborhoods to the HOA changed both the overall population and the number of seats. Since each seat on the HOA Board represented approximately 10 residents, adding 10 more seats for the 102 new residents is a proportional increase. Twin Pines and Ashville rightfully got the new seats they deserved, but the extra seat previously given to Maplewood Acres ended up going to Oakey Oaks.

After going through a few different apportionment methods, Congress finally adopted Hamilton's method in 1852. However, since 1940, the United States House of Representatives has used the **Huntington-Hill Method** for apportionment. Although not covered in this book, more information about the Hill-Huntington Method can be found in the vast Wikipedia archives at http://en.wikipedia.org/wiki/Huntington-Hill_method.

Finally, Mathematicians Michel L. Balinski and H. Peyton Young established **Balinski and Young's Theorem**, which states there is no method of apportionment that satisfies the quota rule and also avoids these paradoxes.

Section 6.5 Exercises

1. A small country with three regions has 25 seats in its legislature. The population of each region is as follows:

   - Eastern = 4,680

   - Central = 2,064

   - Western = 1,431

   a. What is the standard divisor? Round your answer to the nearest thousandth.

   b. What is the quota for the each region? Round your answer to the nearest hundredth.

   c. Using the Hamilton Method, how many seats in the legislature will be apportioned to each region?

2. A state with 13 seats to be apportioned is to be divided into three districts with the following populations:

   - Eastern: 16,280

   - Central: 13,490

   - Western: 9,560

a. What is the standard divisor? Round your answer to the nearest thousandth.

b. What is the quota for the each district? Round your answer to the nearest hundredth.

c. Using the Hamilton Method, how many seats will be apportioned to each district?

3. Three friends have pooled their finances to buy 20 bottles of vintage wine. They decide to divide the bottles using the Hamilton Method based on the amount of money each person contributed.

- Jerod: $295

- Michael: $205

- Rob: $390

a. What is the standard divisor? Round your answer to the nearest thousandth.

b. What is the quota for the each person? Round your answer to the nearest hundredth.

c. How many bottles of wine will be apportioned to each person?

4. A high school government is made up of 50 seats. The populations of the classes are as follows:

   - Seniors = 275

   - Juniors = 767

   - Sophomores = 465

   - Freshmen = 383

   a. What is the standard divisor? Round your answer to the nearest thousandth.

   b. What is the quota for the each class? Round your answer to the nearest hundredth.

   c. Using the Hamilton Method, how many seats will be apportioned to each class?

5. A country has four territories with the populations listed below and currently has 315 seats in the legislature.

   - Northalia has a population of 896.

   - Southville has a population of 426.

   - Eastburgh has a population of 1,166.

   - Westonia has a population of 667.

    a. Using the Hamilton Method, how will those 315 seats be apportioned?

    b. The legislature is changing and one new seat will be added. Using the same populations and the Hamilton Method, how will the new 316-seat legislature be apportioned?

    c. Is there a paradox? If so, what type?

6. A country has the following 4 territories with the populations listed below and currently has 149 seats in the legislature.

- Northalia has a population of 896.

- Southville has a population of 426.

- Eastburgh has a population of 1,166.

- Westonia has a population of 667.

    a. Using the Hamilton Method, how will those 149 seats be apportioned?

    b. The legislature is changing, and one new seat will be added. Using the same populations and the Hamilton Method, how will the new 150-seat legislature be apportioned?

c. Is there a paradox? If so, what type?

7. A country has the following 4 states and 50 seats in the legislature.

	Old Census	New Census
Northalia	27,200	28,102
Southville	18,600	19,307
Eastburgh	11,400	11,404
Westonia	6,250	6,244

a. Using the old census data and Hamilton method, how will the 50 seats be apportioned?

b. Using the new census data and Hamilton method, how will the 50 seats be apportioned?

c. Is there a paradox? If so, what type?

8. A country has the following 4 states and 200 seats in the legislature.

	Old Census	New Census
Northalia	93,700	94,502
Southville	35,850	35,853
Eastburgh	56,850	56,859
Westonia	3,450	3,448

a. Using the old census data and Hamilton Method, how will the 200 seats be apportioned?

b. Using the new census data and Hamilton Method, how will the 200 seats be apportioned?

c. Is there a paradox? If so, what type?

9. A homeowner's association consists of three neighborhoods, with the following resident counts.

	Residents
Rosedale	124
Sunnyville	366
Talon	218

a. Using the Hamilton Method, how will the 20 Board seats be apportioned?

b. If the association votes to add the 54-resident neighborhood of Underwood and two additional seats on the Board, how will the apportionment change?

c. Is there a paradox? If so, what type?

10. The United States House of Representatives uses which of the following methods of apportionment?

    a. Randomly assigning the number of seats given to each state.

    b. The Hamilton Method

    c. The Huntington-Hill Method

    d. None of these

Answers to Section 6.5 Exercises

1. a. 327.000

    b. Eastern = 14.31, Central = 6.31, Western = 4.38

    c. Eastern = 14, Central = 6, Western = 5

2. a. 3,025.385

    b. Eastern = 5.38, Central = 4.46, Western = 3.16

    c. Eastern = 5, Central = 5, Western = 3

3. a. 44.500

    b. Jerod = 6.63, Michael = 4.61, Rob = 8.76

    c. Jerod = 7, Michael = 4, Rob = 9

4. a. 37.800

    b. Seniors = 7.28, Juniors = 20.29, Sophomores = 12.30, Freshmen = 10.13

    c. Seniors = 7, Juniors = 20, Sophomores = 13, Freshmen = 10

5. a. Northalia = 89, Southville = 43, Eastburgh = 116, Westonia = 67

    b. Northalia = 90, Southville = 42, Eastburgh = 117, Westonia = 67

    c. Yes. Alabama Paradox

6. a. Northalia = 42, Southville = 20, Eastburgh = 55, Westonia = 32

    b. Northalia = 43, Southville = 20, Eastburgh = 55, Westonia = 32

    c. There is no paradox.

7. a. Northalia = 21, Southville = 15, Eastburgh = 9, Westonia = 5

   b. Northalia = 21, Southville = 15, Eastburgh = 9, Westonia = 5

   c. There is no paradox.

8. a. Northalia = 99, Southville = 38, Eastburgh = 60, Westonia = 3

   b. Northalia = 99, Southville = 37, Eastburgh = 60, Westonia = 4

   c. Yes. Population Paradox

9. a. Rosedale = 4, Sunnyville = 10, Talon = 6

   b. Rosedale = 4, Sunnyville = 11, Talon = 6, Underwood = 1

   c. Yes. New State Paradox—Sunnyville did nothing, and gained a seat.

10. c

Credits

1. "Perikles ostracon," http://commons.wikimedia.org/wiki/File:Perikles_ostracon.svg. Copyright in the Public Domain.

2. "HAMILTON, Alexander-Treasury (BEP engraved portrait)," http://commons.wikimedia.org/wiki/File:HAMILTON,_Alexander-Treasury_(BEP_engraved_portrait).jpg. Copyright in the Public Domain.

3. "Proposed Electoral College 2012," http://commons.wikimedia.org/wiki/File:Proposed_Electoral_College_2012.svg. Copyright in the Public Domain.

4. "Soft Serve Ice Cream," http://pixabay.com/en/soft-ice-cream-cone-vanilla-snack-617724/. Copyright in the Public Domain.

5. "Vote," http://pixabay.com/en/choice-elect-election-vote-button-158159/. Copyright in the Public Domain.

6. Copyright © Ebayzme (CC BY-SA 3.0) at http://commons.wikimedia.org/wiki/File:Frazerbrown.jpg.

7. "Voting," http://pixabay.com/en/elections-vote-sheet-paper-pen-536656/. Copyright in the Public Domain.

8. "Cartoon Girl Soccer Player," http://pixabay.com/en/girl-soccer-play-seven-black-hair-306824/. Copyright in the Public Domain.

9. "Cartoon Soccer Ball," http://pixabay.com/en/ball-soccer-sports-football-round-295326/. Copyright in the Public Domain.

10. "Baseball Collage," http://pixabay.com/en/cooperstown-baseball-hall-of-fame-109030/. Copyright in the Public Domain.

11. "Questions Mark," http://pixabay.com/en/team-spirit-team-question-mark-437507/. Copyright in the Public Domain.

12. "Cartoon Handshake," http://pixabay.com/en/handshake-connection-friendship-310912/. Copyright in the Public Domain.

13. "Birthday Cake," http://pixabay.com/en/cake-birthday-candle-sweet-dessert-35700/. Copyright in the Public Domain.

14. "iPod Cartoon," http://pixabay.com/en/ipod-apple-mp3-player-gadget-37138/. Copyright in the Public Domain.

15. "Speaker Cartoon," http://pixabay.com/en/speaker-loudspeaker-amplified-35551/. Copyright in the Public Domain.

16. "Baseball," http://pixabay.com/en/baseball-white-red-designs-38208/. Copyright in the Public Domain.

17. "Teddy Bear," http://pixabay.com/en/teddy-bear-sad-plaything-gift-fur-315390/. Copyright in the Public Domain.

18. Cassius Marcellus Coolidge, "A Friend in Need," http://commons.wikimedia.org/wiki/File:A_Friend_in_Need_1903_C.M.Coolidge.jpg. Copyright in the Public Domain.

19. "The Thinker," http://pixabay.com/en/thinker-person-sit-sitting-111253/. Copyright in the Public Domain.

20. "Chocolate Chip Cookie," http://pixabay.com/en/cookie-chocolate-chip-blurred-307960/. Copyright in the Public Domain.

21. "Vintage Car," http://pixabay.com/en/classic-car-car-vintage-purple-152118/. Copyright in the Public Domain.

22. "Playing Cards," http://pixabay.com/en/boys-people-playing-trading-cards-17448/. Copyright in the Public Domain.

23. Vectorportal, "Alabama Vector Map," http://www.vectorportal.com/subcategory/87/ALABAMA-VECTOR-MAP.eps/ifile/3396/detailtest.asp. Copyright © by VectorPortal. Reprinted with permission.

24. "Wine Bottles," http://pixabay.com/en/wine-drink-glass-416051/. Copyright in the Public Domain.

Index

CPSIA information can be obtained
at www.ICGtesting.com
Printed in the USA
FSOW03n0254241215
14897FS

9 781634 872508